Lectures in Mathematics
ETH Zürich
Department of Mathematics
Research Institute of Mathematics

Managing Editor:
Michael Struwe

Frédéric Hélein

Constant Mean Curvature Surfaces, Harmonic Maps and Integrable Systems

Notes taken by Roger Moser

Birkhäuser Verlag
Basel · Boston · Berlin

Author's address:

CMLA
Ecole Normale Supérieure de Cachan
61, avenue du Président Wilson
94235 Cachan Cedex
France

2000 Mathematical Subject Classification 53C42, 53C43; 53C28, 53C35, 70C06

A CIP catalogue record for this book is available from the
Library of Congress, Washington D.C., USA

Deutsche Bibliothek Cataloging-in-Publication Data
Hélein, Frédéric:
Constant mean curvature surfaces, harmonic maps integrable systems /
Frédéric Hélein. Notes taken by Roger Moser. - Basel ; Boston ; Berlin :
Birkhäuser, 2001
   (Lectures in mathematics : ETH Zürich)
   ISBN 3-7643-6576-5

ISBN 3-7643-6576-5 Birkhäuser Verlag, Basel – Boston – Berlin

© 2001 Birkhäuser Verlag, P.O. Box 133, CH-4010 Basel, Switzerland
Member of the BertelsmannSpringer Publishing Group
Printed on acid-free paper produced from chlorine-free pulp. TCF ∞

ISBN 3-7643-6576-5

9 8 7 6 5 4 3 2 1

http://www.birkhauser.ch

Contents

Preface

One of the most striking development of the last decades in the study of minimal surfaces, constant mean surfaces and harmonic maps is the discovery that many classical problems in differential geometry – including these examples – are actually integrable systems.

This theory grew up mainly after the important discovery of the properties of the Korteweg-de Vries equation in the sixties. After C. Gardner, J. Greene, M. Kruskal et R. Miura [44] showed that this equation could be solved using the inverse scattering method and P. Lax [62] reinterpreted this method by his famous equation, many other deep observations have been made during the seventies, mainly by the Russian and the Japanese schools. In particular this theory was shown to be strongly connected with methods from algebraic geometry (S. Novikov, V.B. Matveev, I.M. Krichever...), loop techniques (M. Adler, B. Kostant, W.W. Symes, M.J. Ablowitz...) and Grassmannian manifolds in Hilbert spaces (M. Sato...). Approximatively during the same period, the twistor theory of R. Penrose, built independentely, was applied successfully by R. Penrose and R.S. Ward for constructing self-dual Yang-Mills connections and four-dimensional self-dual manifolds using complex geometry methods. Then in the eighties it became clear that all these methods share the same roots and that other instances of integrable systems should exist, in particular in differential geometry. This led K. Uhlenbeck [82] to describe harmonic maps on a two-sphere, with values in $U(n)$ using families of curvature free connections depending on a complex "spectral" parameter. At the same period N. Hitchin [54] investigated finite type tori into $SU(2)$ starting from similar methods. Such formulations were already proposed in the seventies by K. Pohlmeyer [69], V.E. Zhakarov-A.V. Mikhailov [91] and V.E. Zhakarov-A.B. Shabat [92]. A catalysor of these developments was the construction by H. Wente in 1984 of an immersed constant mean curvature torus in $\mathbb{R}^3$, which had the effect of removing an old inhibition due to the fact that people believed that such tori should not exist. Many results then followed quickly and we have now a very rich and fruitful theory for constructing constant mean curvature surfaces and harmonic maps of surfaces with values into symmetric manifolds using integrable systems methods.

All that seems apparently a new theory, but many features of the "completely integrable behaviour" of constant mean surfaces have been guessed by geometers of the nineteenth century. Namely the existence of associated families of such surfaces by O. Bonnet, the study of special surfaces with planar curvature lines by A. Enneper and his students and the various Bäcklund transformations discovered and studied by A.V. Bäcklund, L. Bianchi, S. Lie, G. Darboux, E. Goursat and J. Clairin (see [71] for details and references).

This Monograph is intended to give an introduction to this old and new theory from the point of view of differential geometry. For that reason, it has seemed more natural for me to introduce the existence of families of curvature free connections for harmonic maps starting from the associated family of constant mean surfaces. Note however that this is not the historical way the theory actually developed and contemporary people were initially more inspired by the example of the KdV equation and the twistor theory. We also presented here some basic exposition of the twistor theory for harmonic maps, which was initiated to my knowledge by E. Calabi. Indeed this theory was an important stimulation for the integrable system theory, in particular in the work of K. Uhlenbeck, and shares some similarities with the integrable systems theory (for a complete exposition, see [23]). I also made an effort to present the beautiful result of U. Pinkall and I. Sterling concerning constant mean curvature tori (Chapter 9) in the framework of loop groups theory, in order to show how this result connects with the rest of theory, which uses loop groups.

The present text is just an introduction and is far from being complete. We may recommend as parallel lectures the books [46] and [43] plus of course . . . reading the cited papers.

These Notes come from a lecture that I gave at the Eidgenössische Technische Hochschule Zürich during Spring 1999. Most parts of the text were written and typed by R. Moser. I wish here to thank the ETH Zürich and more particularly Prof. M. Struwe for his hospitality and to thank R. Moser for his very nice work.

I also want to thank Joseph Dorfmeister and Pascal Romon for their valuable comments on this text.

1 Introduction: Surfaces with prescribed mean curvature

Curvature

For a curve Γ in a plane, and for any point on this curve, there exists a circle (or a straight line) which is the best approximation at this point of the curve up to third order. The inverse of its radius $k = \frac{1}{R}$ (or $k = 0$ if the best approximation is a straight line) is called the curvature of the curve at the given point.

For a surface Σ in $\mathbb{R}^3$ and a point $m \in \Sigma$, we consider the 1-parameter family of affine planes which contain the straight line passing through m and being normal to Σ at m. Each of these planes locally intersects the surface along a planar curve containing m. We may label these planes by choosing one, say P_0, and for $\theta \in \mathbb{R}/2\pi\mathbb{Z}$ naming by P_θ the image of P_0 by a rotation of angle θ around the normal line to Σ at m. Then the curvature $k(\theta)$ of $P_\theta \cap \Sigma$ at m is of the form $\frac{k_1+k_2}{2} + \frac{k_1-k_2}{2}\cos(2(\theta - \theta_0))$, where k_1 and k_2 are two numbers, called the principal curvatures of Σ at m. They are the extremal curvatures, achieved at two extremal positions of P_θ which are orthogonal (the principal directions). The average $H = \frac{1}{2}(k_1 + k_2)$ of k_1 and k_2 is called the mean curvature of Σ at m, their product $K = k_1 k_2$ the Gauss curvature.

If $k_1 = k_2$, i. e. if the curvature $k(\theta)$ is a constant, then the point m is called umbilic.

Experiments with soap

We are going to describe some experiments, carried out by J. Plateau, which may give some physical motivation for what is to follow.

Imagine a piece of wire, bent in the shape of a closed curve Γ, that is dipped into a solution of water and soap, such that it becomes spanned with a soap film. This film will take the form of a surface Σ_0 with boundary $\partial\Sigma_0 = \Gamma$, having least area among all surfaces with the same boundary. Computing the Euler-Lagrange equation of this variation problem, we see that Σ_0 satisfies

$$H = 0, \tag{1.1}$$

where H is the mean curvature of Σ_0 as described above. Such a surface is called a *minimal surface*.

Suppose now that there is some device that allows to have different pressure on either side of the soap film in question. Then instead of (1.1), the equation

$$H = C \tag{1.2}$$

will hold for Σ_0, where C is a constant different from 0 (depending on the difference of the pressures applied). In this case we call Σ_0 a *constant mean curvature* (CMC) *surface*.

First and second fundamental form

Let Σ be a piece of a surface, embedded in $\mathbb{R}^3$, which is diffeomorphic to the unit disk $D^2 = \{z = x + iy \in \mathbb{C} \colon |z| < 1\}$. Let $X \colon D^2 \to \mathbb{R}^3$ be a parametrization of Σ. Then we can define the following.

Definition 1.1 *The first fundamental form of an embedding $X \colon D^2 \to \mathbb{R}^3$ is the quadratic form given by the matrix*

$$
\mathrm{I} = \begin{pmatrix} \left| \frac{\partial X}{\partial x} \right|^2 & \left\langle \frac{\partial X}{\partial x}, \frac{\partial X}{\partial y} \right\rangle \\ \left\langle \frac{\partial X}{\partial x}, \frac{\partial X}{\partial y} \right\rangle & \left| \frac{\partial X}{\partial y} \right|^2 \end{pmatrix},
$$

which depends on $z \in D^2$.

For $\xi = (\xi_1, \xi_2) \in \mathbb{R}^2$, we compute ${}^t\xi\, \mathrm{I}\, \xi = |dX(\xi)|^2$. Now fix an orientation of Σ by assuming that $(\frac{\partial X}{\partial x}, \frac{\partial X}{\partial y})$ is an oriented basis of $T_{X(z)}\Sigma$. Consider

$$
u = \frac{\frac{\partial X}{\partial x} \times \frac{\partial X}{\partial y}}{\left| \frac{\partial X}{\partial x} \times \frac{\partial X}{\partial y} \right|},
$$

where $\times$ denotes the vector product in $\mathbb{R}^3$. The map $u \colon D^2 \to S^2$ thus given is called the Gauss map of X.

Definition 1.2 *The second fundamental form of X is the quadratic form given by*

$$
\mathrm{II} = \begin{pmatrix} \left\langle \frac{\partial^2 X}{\partial x^2}, u \right\rangle & \left\langle \frac{\partial^2 X}{\partial x \partial y}, u \right\rangle \\ \left\langle \frac{\partial^2 X}{\partial x \partial y}, u \right\rangle & \left\langle \frac{\partial^2 X}{\partial y^2}, u \right\rangle \end{pmatrix}.
$$

With the notions of the first and second fundamental form, we are able to define the principal curvatures of a surface in a more convenient way than before.

Definition 1.3 *The principal curvatures k_1, k_2 are the eigenvalues of II with respect to I, i. e. the solutions of*

$$
\det(\mathrm{II} - \lambda \mathrm{I}) = 0.
$$

By a multiplication of $II - \lambda I$ with I^{-1} from the right it is easily verified that the mean curvature and the Gauss curvature can be computed from I and II in the following way:

$$H = \frac{1}{2}\text{tr}(II \cdot I^{-1}), \quad K = \det(II)/\det(I). \tag{1.3}$$

Conformal coordinates

To simplify things, we will use conformal coordinates. Choose Σ like above. The following is a well-known result.

Theorem 1.1 *There exists an embedding map $X: D^2 \to \mathbb{R}^3$, such that*

i) $X(D^2) = \Sigma$, and

ii) X is conformal, i. e.

$$\left|\frac{\partial X}{\partial x}\right|^2 - \left|\frac{\partial X}{\partial y}\right|^2 = \left\langle \frac{\partial X}{\partial x}, \frac{\partial X}{\partial y} \right\rangle = 0. \tag{1.4}$$

Note that (1.4) means that the vectors $\frac{\partial X}{\partial x}$ and $\frac{\partial X}{\partial y}$ have at each point the same length and are perpendicular to each other. This implies that there is a function $\omega: D^2 \to \mathbb{R}$, such that

$$\frac{\partial X}{\partial x} = e^\omega e_1, \quad \frac{\partial X}{\partial y} = e^\omega e_2,$$

and (e_1, e_2) establish an orthonormal basis of $T_{X(z)}\Sigma$ for each $z \in D^2$. Thus we may write

$$I = e^{2\omega}\begin{pmatrix} 1 & 0 \\ 0 & 1 \end{pmatrix}, \quad II = e^{2\omega}\begin{pmatrix} h_{11} & h_{21} \\ h_{12} & h_{22} \end{pmatrix}.$$

Then k_1 and k_2 are the solutions of

$$\det\left(\begin{pmatrix} h_{11} & h_{21} \\ h_{12} & h_{22} \end{pmatrix} - \lambda\begin{pmatrix} 1 & 0 \\ 0 & 1 \end{pmatrix}\right) = 0,$$

i. e. the eigenvalues of the matrix

$$\begin{pmatrix} h_{11} & h_{21} \\ h_{12} & h_{22} \end{pmatrix}.$$

Consequently,

$$H = \frac{h_{11} + h_{22}}{2}, \quad K = \det\begin{pmatrix} h_{11} & h_{21} \\ h_{12} & h_{22} \end{pmatrix}.$$

We see that, for conformal X,

$$2H = h_{11} + h_{22} = e^{-2\omega} \langle \Delta X, u \rangle,$$

or

$$\langle \Delta X, u \rangle = 2e^{2\omega} H.$$

Moreover, we have the following.

Lemma 1.1 *If X is conformal, then*

$$\Delta X \perp \frac{\partial X}{\partial x}, \frac{\partial X}{\partial y}.$$

This can easily be verified by differentiating (1.4). The details are left to the reader. It allows, however, the following conclusions.

Corollary 1.1 *The map X satisfies*

$$\Delta X = \langle \Delta X, u \rangle u = 2He^{2\omega}u = 2H \frac{\partial X}{\partial x} \times \frac{\partial X}{\partial y},$$

as soon as it is conformal.

Apply this to the equation (1.2) in the form

$$H = C = H_0.$$

We see that it is equivalent to

$$\Delta X = 2H_0 \frac{\partial X}{\partial x} \times \frac{\partial X}{\partial y}.$$

In particular this implies that minimal surfaces are always images of harmonic and conformal parametrizations. This leads to the problem of finding maps $X: D^2 \to \mathbb{R}^3$, such that

- X is conformal, i. e.

$$0 = \left| \frac{\partial X}{\partial x} \right|^2 - \left| \frac{\partial X}{\partial y} \right|^2 - 2i \left\langle \frac{\partial X}{\partial x}, \frac{\partial X}{\partial y} \right\rangle,$$

- $\Delta X = 0$.

Weierstrass representation

Using the notation of complex differentiation,

$$\frac{\partial}{\partial z} = \frac{1}{2}\left(\frac{\partial}{\partial x} - i\frac{\partial}{\partial y}\right), \quad \frac{\partial}{\partial \bar{z}} = \frac{1}{2}\left(\frac{\partial}{\partial x} + i\frac{\partial}{\partial y}\right), \quad \Delta = 4\frac{\partial^2}{\partial z\partial\bar{z}},$$

we find that our problem can be expressed in the form

$$\left\{ \begin{array}{rcl} \left(\dfrac{\partial X}{\partial z}\right)^2 & = & 0, \\[2ex] \dfrac{\partial}{\partial\bar{z}}\left(\dfrac{\partial X}{\partial z}\right) & = & 0. \end{array} \right.$$

Write $f = 2\frac{\partial X}{\partial z} \colon D^2 \to \mathbb{C}^3$. We want to solve

$$f^2 = 0, \quad \frac{\partial f}{\partial\bar{z}} = 0.$$

This can be done explicitly by the representation

$$f = \begin{pmatrix} \frac{i}{2}(w^2 - 1) \\ \frac{1}{2}(w^2 + 1) \\ iw \end{pmatrix} h,$$

where $w, h \colon D^2 \to \mathbb{C}$ are holomorphic functions. (The function w might be meromorphic.) Eventually, this gives us Weierstrass representation of a solution of our problem:

$$X(z) = \mathrm{Re}\left[\int_{z_0}^{z} f(\zeta)\, d\zeta\right],$$

where z_0 is an arbitrarily chosen point in D^2.

Completely integrable systems

The Weierstrass representation for minimal surfaces seems to be a miracle, since it describes all solutions of a nonlinear geometrical problem by a very simple algebraic construction. We shall see in these lectures more sophisticated miracles, occuring in various geometrical situations. People called them *completely integrable systems*. Classically, they are nonlinear equations on which mathematicians and physicists discovered unusual properties. Some of these properties are

- **existence of solitons.** This terminology comes from evolution problems and the most famous example is the Korteweg-de Vries (or KdV) equation

$\frac{\partial u}{\partial t} + 6u\frac{\partial u}{\partial x} + \frac{\partial^3 u}{\partial x^3} = 0$, modelling water waves in a flat shallow channel. The story started in 1834 with the experimental observation of a solitary wave travelling along such a channel along a very long distance by J.S. Russel [73]. A model equation was derived in 1895 by D.J. Korteweg and G. de Vries [59]. Solitons are solutions of nonlinear partial differential equations which are localised in space, whose profile is not dispersed after a long period of time and which resist to interactions with other solitons [1]. Thus it is a smooth field, solution of some partial differential equation, which behaves like a particle.

- **Bäcklund transformations.** A baby example is the following: start from a harmonic function f of two real variables x and y. We may write the Laplace equation for f as $d\left(-\frac{\partial f}{\partial y}dx + \frac{\partial f}{\partial x}dy\right) = 0$, which implies that, if we work on a simply connected domain, there exists a function g such that $-\frac{\partial f}{\partial y}dx + \frac{\partial f}{\partial x}dy = dg$. Then g is another harmonic function, namely the conjugate function of f (i.e. $f+ig$ is a holomorphic function of $x+iy$). Such transformations, producing a solution of some partial differential equation starting from another solution work also in nonlinear situations.

- *by a nonlinear change of variable, the problem reduces to solving linear equations.* Weierstrass representation does it obviously, reducing the minimal surface equation to the Cauchy-Riemann system. We shall meet the same situation, but involving a much more complicated "change of variable".

- **a Hamiltonian structure.** Solutions are spanned by the Hamiltonian flows of commuting functions on a symplectic (or Poisson) manifold. This is described by Liouville's theorem.

- **Infinitely many symmetries.** Infinite dimensional Lie groups acts on the set of solutions of these systems.

The theory of completely integrable systems is not a pragmatic point of view, in the sense that it does not really provide a method (like the analytic approach based on functional analysis) where, starting from some qualitative intuition, one works (often hardly) to construct the tools one needs for proving (or disproving) what one believe to be true (existence, regularity,...). Indeed working on completely integrable systems is rather based on a contemplation of some very exceptional equations which hide a Platonic structure: although these equations do not look trivial a priori, we shall discover that they are elementary, once we understand how they are encoded in the language of symplectic geometry, Lie groups and algebraic geometry. It will turn out that this contemplation is fruitful and leads to many results.

---

[1]for the KdV equation, the simplest solution involving the soliton behaviour is the one-soliton $u(x,t) = 2a^2\text{sech}^2(a(x - 4a^2 t))$, for $a > 0$

2 From minimal surfaces and CMC surfaces to harmonic maps

Let Ω be an open subset of $\mathbb{C}$ and $X\colon\Omega\to\mathbb{R}^3$ a conformal parametrization of a surface Σ. We use all the notation form the previous section. In particular, we have for the mean curvature

$$H = \frac{h_{11} + h_{22}}{2},$$

if we write for the second fundamental form

$$\mathrm{II} = e^{2\omega}\left(\begin{array}{cc} h_{11} & h_{21} \\ h_{12} & h_{22} \end{array}\right).$$

It is clear that II is symmetric. So, for a fixed H, there exist functions $a, b\colon\Omega\to\mathbb{R}$ such that

$$\left(\begin{array}{cc} h_{11} & h_{21} \\ h_{12} & h_{22} \end{array}\right) = \left(\begin{array}{cc} H+a & b \\ b & H-a \end{array}\right). \tag{2.1}$$

For convenience, we denote coordinates on $\mathbb{C}$ alternatively (x, y) or (x^1, x^2), where $x = x^1$, $y = x^2$. Because of the fact

$$\left\langle \frac{\partial u}{\partial x^i}, \frac{\partial X}{\partial x^j} \right\rangle + \left\langle u, \frac{\partial^2 X}{\partial x^i \partial x^j} \right\rangle = 0, \quad i, j = 1, 2,$$

which is a consequence of the orthogonality of u to $\frac{\partial X}{\partial x}$ and $\frac{\partial X}{\partial y}$, we can write II in yet another form:

$$\mathrm{II} = -\left(\left\langle \frac{\partial u}{\partial x^i}, \frac{\partial X}{\partial x^j} \right\rangle\right)_{i,j=1,2}. \tag{2.2}$$

Note furthermore that $\langle u, \frac{\partial u}{\partial x^i} \rangle = 0$ for $i = 1, 2$ (because of $|u| = 1$).

The map X provides a moving frame on the surface Σ, i. e. three vector fields that constitute a orthonormal and oriented basis of $\mathbb{R}^3$ at each point of Σ. It consists of the vector fields

$$e_1 = e^{-\omega}\frac{\partial X}{\partial x}, \quad e_2 = e^{-\omega}\frac{\partial X}{\partial y}, \quad u = e_1 \times e_2.$$

Thus

$$\frac{\partial u}{\partial x^i} = e^{-\omega}\left\langle \frac{\partial u}{\partial x^i}, \frac{\partial X}{\partial x} \right\rangle e_1 + e^{-\omega}\left\langle \frac{\partial u}{\partial x^i}, \frac{\partial X}{\partial y} \right\rangle e_2, \quad i = 1, 2.$$

Using (2.1) and (2.2), we finally obtain

$$\left(\frac{\partial u}{\partial x}, \frac{\partial u}{\partial y}\right) = -\left(\frac{\partial X}{\partial x}, \frac{\partial X}{\partial y}\right)\left(\begin{array}{cc} H+a & b \\ b & H-a \end{array}\right). \tag{2.3}$$

2.1   Minimal surfaces

Suppose now that Σ is a minimal surface. Then

$$\left(\frac{\partial u}{\partial x}, \frac{\partial u}{\partial y}\right) = -\left(\frac{\partial X}{\partial x}, \frac{\partial X}{\partial y}\right)\left(\begin{array}{cc} a & b \\ b & -a \end{array}\right). \qquad (2.4)$$

Comparison with antiholomorphic functions

A function $f: \Omega \to \mathbb{C}$ is defined to be antiholomorphic, if

$$\frac{\partial f}{\partial z} = \frac{\partial f}{\partial x} - i\frac{\partial f}{\partial y} = 0.$$

In other terms,

$$\left(\frac{\partial f}{\partial x}, \frac{\partial f}{\partial y}\right) = (1, i)\left(\begin{array}{cc} a & b \\ b & -a \end{array}\right),$$

if a denotes the real part and b the imaginary part of $\frac{\partial f}{\partial x}$. This bears some resemblance to (2.4). We will see that this is no coincidence. As a matter of fact, u can be seen as an antiholomorphic function in a certain way.

Another point of view on antiholomorphic functions comes form the following (equivalent) definition.

Definition 2.1 *A map $f: \Omega \to \mathbb{C}$ is antiholomorphic if and only if*

i) it is weakly conformal, i. e.

$$\left|\frac{\partial f}{\partial x}\right|^2 - \left|\frac{\partial f}{\partial y}\right|^2 - 2i\left\langle\frac{\partial f}{\partial x}, \frac{\partial f}{\partial y}\right\rangle = 0,$$

and

ii) it reverses orientation, i. e.

$$\det\left(\frac{\partial f}{\partial x}, \frac{\partial f}{\partial y}\right) \leq 0.$$

The difference between condition i) and the definition for conformal maps is that weakly conformal maps are allowed to have vanishing derivatives.

Now note that (e_1, e_2) is an oriented basis of $T_u S^2$ at each point in the sense that it constitutes an oriented basis of $\mathbb{R}^3$ together with u. Thus (2.4) shows that u is both weakly conformal and orientation reversing. On the other hand,

S^2 can be mapped to $\mathbb{C} \cup \{\infty\}$ by a conformal and orientation preserving map, the stereographic projection

$$P \colon S^2 \to \mathbb{C} \cup \{\infty\}, \quad (x^1, x^2, x^3) \mapsto \frac{x^1 + ix^2}{1 + x^3}$$

which has the inverse map

$$P^{-1}(z) = \frac{1}{1 + |z|^2}(z + \bar{z}, -i(z - \bar{z}), 1 - |z|^2).$$

The composition $P \circ u \colon \Omega \to \mathbb{C} \cup \{\infty\}$ is thus weakly conformal and orientation reversing, hence antiholomorphic on the set $\Omega \backslash u^{-1}\{(0, 0, -1)\}$. This means

$$\frac{\partial}{\partial z} P \circ u = 0.$$

Set $v = -\frac{1}{\overline{P \circ u}} = P \circ (-u)$. This function is clearly meromorphic. Furthermore, it allows the representation

$$u = -P^{-1} \circ v = -\frac{1}{1 + |v|^2} \begin{pmatrix} v + \bar{v} \\ -i(v - \bar{v}) \\ 1 - |v|^2 \end{pmatrix}.$$

Now consider again (2.4). In complex notation it is

$$\frac{\partial u}{\partial \bar{z}} = -(a + ib)\frac{\partial X}{\partial z}.$$

So by differentiation we obtain

$$\begin{aligned}
\frac{\partial X}{\partial z} &= \frac{1}{a + ib}\frac{\partial}{\partial \bar{z}}\left(\frac{1}{1 + |v|^2}\begin{pmatrix} v + \bar{v} \\ -i(v - \bar{v}) \\ 1 - |v|^2 \end{pmatrix}\right) \\
&= \frac{2i}{(a + ib)(1 + |v|^2)^2}\frac{\partial \bar{v}}{\partial \bar{z}}\begin{pmatrix} \frac{i}{2}(v^2 - 1) \\ \frac{1}{2}(v^2 + 1) \\ iv \end{pmatrix},
\end{aligned}$$

which corresponds to Weierstrass representation.

Harmonic maps

We are going to exploit (2.4) some more. Take the vector product with u to get

$$\begin{aligned}
u \times \frac{\partial u}{\partial x} &= -\left(a\frac{\partial X}{\partial y} - b\frac{\partial X}{\partial x}\right) = -\frac{\partial u}{\partial y}, \\
u \times \frac{\partial u}{\partial y} &= -\left(a\frac{\partial X}{\partial x} + b\frac{\partial X}{\partial y}\right) = \frac{\partial u}{\partial x}.
\end{aligned}$$

As a consequence of this we have

$$u \times \Delta u = \frac{\partial}{\partial x}\left(u \times \frac{\partial u}{\partial x}\right) + \frac{\partial}{\partial y}\left(u \times \frac{\partial u}{\partial y}\right) = -\frac{\partial^2 u}{\partial x \partial y} + \frac{\partial^2 u}{\partial y \partial x} = 0,$$

which means that $\Delta u \perp T_u S^2$. This property is worth a little consideration.

Definition 2.2 *A map $u \colon \Omega \to S^2$ is called harmonic if and only if it is a solution of $\Delta u \perp T_u S^2$. More generally, let $\mathcal{N}$ be a submanifold of dimension n without boundary of an Euclidean space $\mathbb{R}^N$. Let Ω be an open subset of $\mathbb{R}^m$. A map $u \colon \Omega \to \mathcal{N}$ is called harmonic if and only if $\Delta u \perp T_u \mathcal{N}$ everywhere.*

For instance, if Ω is an open interval (a, b), then a curve $\gamma \colon (a, b) \to \mathcal{N}$ is harmonic if and only if it satisfies $\ddot{\gamma}(t) \perp T_{\gamma(t)} \mathcal{N}$, which means that γ is a constant speed parametrization of a geodesic. Or, if $\mathcal{N} = \mathbb{R}$, then the harmonic maps $\Omega \to \mathbb{R}$ are just the solutions of the Laplace equation $\Delta u = 0$.

2.2   Constant mean curvature surfaces

Now let's see what happens for CMC surfaces. In this case, instead of (2.4) we have to use (2.3). It implies

$$\begin{aligned}
u \times \frac{\partial u}{\partial x} &= -(H+a)\frac{\partial X}{\partial y} + b\frac{\partial X}{\partial x} = -\frac{\partial u}{\partial y} - 2H\frac{\partial X}{\partial y}, \\
u \times \frac{\partial u}{\partial y} &= (H-a)\frac{\partial X}{\partial x} - b\frac{\partial X}{\partial y} = \frac{\partial u}{\partial x} + 2H\frac{\partial X}{\partial x}.
\end{aligned}$$

Like before we compute

$$u \times \Delta u = \frac{\partial}{\partial x}\left(u \times \frac{\partial u}{\partial x}\right) + \frac{\partial}{\partial y}\left(u \times \frac{\partial u}{\partial y}\right) = \left(-\frac{\partial^2}{\partial x \partial y} + \frac{\partial^2}{\partial y \partial x}\right)(u + 2HX) = 0.$$

Hence also in this case u is harmonic. Thus we have proved half of the following, which is due to E. A. Ruh and J. Vilms, [72].

Theorem 2.1 *Let $X \colon \Omega \to \mathbb{R}^3$ be a conformal immersion of a surface, then its Gauss map $u \colon \Omega \to S^2$ is a harmonic map if and only if the image of X is a CMC surface.*

Converse result

We will see a kind of converse: Any harmonic map $u \colon \Omega \to S^2$ from a simply connected open set Ω in $\mathbb{C}$ into S^2 is the Gauss map of two weakly conformal CMC immersions, both having mean curvature $H = \frac{1}{2}$. First we write the harmonic map condition in a form that is more suitable for this purpose. Since $u \in S^2$, the condition

$$\Delta u \perp T_u S^2$$

is equivalent to

$$0 = u \times \Delta u = \frac{\partial}{\partial x}\left(u \times \frac{\partial u}{\partial x}\right) + \frac{\partial}{\partial y}\left(u \times \frac{\partial u}{\partial y}\right),$$

which implies that the 1-form $u \times \frac{\partial u}{\partial y} dx - u \times \frac{\partial u}{\partial x} dy$ is closed, i. e.

$$d\left(u \times \frac{\partial u}{\partial y} dx - u \times \frac{\partial u}{\partial x} dy\right) = 0.$$

But since Ω is simply connected, this 1-form is even exact. So we can find a map

$$B \colon \Omega \to \mathbb{R}^3$$

that satisfies

$$\begin{cases} \dfrac{\partial B}{\partial x} &= u \times \dfrac{\partial u}{\partial y}, \\[2mm] \dfrac{\partial B}{\partial y} &= -u \times \dfrac{\partial u}{\partial x}. \end{cases} \tag{2.5}$$

This map will help us construct the CMC surfaces we are looking for. But first we have to check a few properties of B.

i) **If B is an immersion (which is true if and only if u is so), then its Gauss curvature is 1.**

To check this, note that the vectors $-\frac{\partial B}{\partial y}$ and $\frac{\partial B}{\partial x}$ come from $\frac{\partial u}{\partial x}$ and $\frac{\partial u}{\partial y}$ by a rotation around u by an angle of $\frac{\pi}{2}$. It is clear therefore that

$$\frac{\partial B}{\partial x} \times \frac{\partial B}{\partial y} = \frac{\partial u}{\partial x} \times \frac{\partial u}{\partial y}.$$

Thus u is the Gauss map of B. Now we use a geometric characterization of the Gauss curvature of a surface: it is the ratio between the pull-back of the area form on S^2 by the Gauss map and the pull-back of the area form on the surface [2]. So we have

$$K = \frac{\left\langle u, \frac{\partial u}{\partial x} \times \frac{\partial u}{\partial y} \right\rangle}{\left\langle u, \frac{\partial B}{\partial x} \times \frac{\partial B}{\partial y} \right\rangle} = 1.$$

Hence, if B is an immersion, its image has a constant Gauss curvature equal to 1. This means that the metric induced by this immersion makes $B(\Omega)$ intrinsically isometric to a subset of S^2, or in other words, that $B(\Omega)$ is a (non-canonical) isometric immersion of the sphere.

ii) **Construction of $X^{\pm}$**

Also by differentiating (2.5), one can prove that

$$\Delta B = 2\left(\frac{\partial u}{\partial x} \times \frac{\partial u}{\partial y}\right) = 2\left(\frac{\partial B}{\partial x} \times \frac{\partial B}{\partial y}\right) = \frac{\partial u}{\partial x} \times \frac{\partial u}{\partial y} + \frac{\partial B}{\partial x} \times \frac{\partial B}{\partial y}. \quad (2.6)$$

Hence B is a harmonic map onto its image. As mentioned earlier, we may see B as a harmonic map into a subset of S^2, immersed isometrically in a non-canonical way. Actually, B is a kind of harmonic conjugate map to u into the sphere.

We want to compute $\Delta(B + u)$ and $\Delta(B - u)$. For u, we have

$$0 = \Delta(|u|^2) = 2\langle u, \Delta u \rangle + 2|du|^2,$$

where $|du|^2 = \left|\frac{\partial u}{\partial x}\right|^2 + \left|\frac{\partial u}{\partial x}\right|^2$. Since we know that Δu is parallel to u, we conclude that

$$\begin{aligned}
\Delta u &= -u|du|^2 \\
&= -\frac{\partial u}{\partial x} \times \left(u \times \frac{\partial u}{\partial x}\right) - \frac{\partial u}{\partial y} \times \left(u \times \frac{\partial u}{\partial y}\right) \qquad (2.7) \\
&= \frac{\partial u}{\partial x} \times \frac{\partial B}{\partial y} + \frac{\partial B}{\partial x} \times \frac{\partial u}{\partial y}.
\end{aligned}$$

The equations (2.6) and (2.7) imply

$$\Delta(B + u) = \frac{\partial(B + u)}{\partial x} \times \frac{\partial(B + u)}{\partial y}, \quad \Delta(B - u) = \frac{\partial(B - u)}{\partial x} \times \frac{\partial(B - u)}{\partial y}.$$

---

[2]Indeed, since $I = {}^t dB.dB$ and $II = -{}^t dB.du$, we deduce that $K = \det(II)/\det(I) = \det({}^t dB.du)/\det({}^t dB.dB)$. But recall the following algebraic fact: if $M = (M_1, M_2)$ and $N = (N_1, N_2)$ are two 2×3 matrices (M_1, M_2, N_1, N_2 being column vectors in $\mathbb{R}^3$), then $\det({}^t M.N) = \langle M_1 \times M_2, N_1 \times N_2 \rangle$. Applying this here, it gives

$$K = \frac{\left\langle \frac{\partial B}{\partial x} \times \frac{\partial B}{\partial y}, \frac{\partial u}{\partial x} \times \frac{\partial u}{\partial y} \right\rangle}{\left|\frac{\partial B}{\partial x} \times \frac{\partial B}{\partial y}\right|^2} = \frac{\left\langle u, \frac{\partial u}{\partial x} \times \frac{\partial u}{\partial y} \right\rangle}{\left\langle u, \frac{\partial B}{\partial x} \times \frac{\partial B}{\partial y} \right\rangle}.$$

Set $X^{\pm} = B \pm u$. Then these maps satisfy

$$\Delta X^{\pm} = \frac{\partial X^{\pm}}{\partial x} \times \frac{\partial X^{\pm}}{\partial y}.$$

iii) **The maps $X^{\pm}$ are weakly conformal.**

This can be checked by a mere calculation. Remember that $-\frac{\partial B}{\partial y}$ and $\frac{\partial B}{\partial x}$ are obtained from $\frac{\partial u}{\partial x}$ and $\frac{\partial u}{\partial y}$ by a rotation by $\frac{\pi}{2}$.

iv) **Conclusion**

We find that X^{+} and X^{-} are both CMC immersions with mean curvature $H = \frac{1}{2}$ (cf. corollary 1.1). The images of X^{+} and X^{-} are parallel to the image of B and pointwise at distance 1 from B. Of course they also have u as their Gauss map. This is a result of O. Bonnet [17] (see also [48]).

The notion of harmonic map can be extended by replacing S^2 by any Riemannian manifold (see the next Chapter), like for instance the n-dimensional sphere S^n. One may then ask whether one can associate to a harmonic map $u : \Omega \subset \mathbb{C} \longrightarrow S^n$ a conformal immersion in a similar fashion as above. This is indeed the case: we can associate to u a conformal immersion X into the Lie algebra $so(n+2)$, unique up to the addition of a constant in $so(n+2)$, by a construction generalizing the above one. For $n = 2$, because of the decomposition $so(4) \simeq so(3) \oplus so(3)$, this immersion eventually splits into two components X^{+} and X^{-}. However the geometric interpretation of X in the general case is less clear: it is related to a kind of Wess-Zumino-Witten problem. For more details, see [48].

Exercise. Let $u : \Omega \to S^2$ be a harmonic map and $B : \Omega \to \mathbb{R}^3$ defined by (2.5). Consider the map $P : \Omega \to \mathbb{R}^3$ defined by

$$P = u \times B.$$

Prove that

i) $P(z) \perp u(z)$ everywhere.

ii) P is a solution of

$$\Delta P + |du|^2 P = 0.$$

iii) For any smooth one-parameter family $u_t : \Omega \to S^2$ of harmonic maps, the map $\frac{\partial u_t}{\partial t} : \Omega \to \mathbb{R}^3$ is called a Jacobi field. Conclude that P is a Jacobi field.

3 Variational point of view and Noether's theorem

Harmonic maps

Consider the set $\mathcal{E}$ consisting of all smooth maps $u\colon \Omega \to \mathcal{N}$, where Ω is an open subset of $\mathbb{R}^m$ and $\mathcal{N}$ a smooth compact submanifold of $\mathbb{R}^N$ without boundary of dimension n. For any $u \in \mathcal{E}$, we define its energy to be

$$E(u) = \int_\Omega \frac{|du|^2}{2}\, dx,$$

where $|du|^2 = \sum_{\alpha=1}^m \left|\frac{\partial u}{\partial x^\alpha}\right|^2$ and $dx = dx^1 \ldots dx^m$. We are interested in maps u that have finite energy (what we will assume subsequently).

More generally, we could equip Ω with a Riemannian metric $g_{\alpha\beta}$. The energy would then be defined similarly, except that instead of dx we'd have $\sqrt{\det(g_{\alpha\beta})}\, dx$ and instead of $|du|^2$ the expression

$$\sum_{\alpha,\beta=1}^m g^{\alpha\beta}(x) \left\langle \frac{\partial u}{\partial x^\alpha}, \frac{\partial u}{\partial x^\beta} \right\rangle,$$

where $(g^{\alpha\beta})$ is the inverse of the matrix $(g_{\alpha\beta})$. We will however not pursue this any more, but restrict our attention to the flat case.

We claim that harmonic maps in the sense of definition 2.2, i. e. solutions $u \in \mathcal{E}$ of

$$\Delta u \perp T_u \mathcal{N},$$

are critical points of the energy functional E in $\mathcal{E}$. To prove this, let $V_\delta \mathcal{N}$ be a tubular neighbourhood of $\mathcal{N}$ of width δ. If δ is sufficiently small, then for any $y \in V_\delta \mathcal{N}$ there exists a unique nearest point to y in $\mathcal{N}$. Thus we can define the nearest point projection

$$\Pi\colon V_\delta \mathcal{N} \to \mathcal{N},$$

which assigns that point to y. Moreover, this map is smooth. Now let $u \in \mathcal{E}$ and $\phi \in C_c^\infty(\Omega, \mathbb{R}^N)$. The map $u_t = \Pi(u + t\phi)$ belongs also to $\mathcal{E}$, provided that $|t|$ is sufficiently small. We are now going to define what exactly is meant by a critical point of E.

Definition 3.1 *The map $u \in \mathcal{E}$ is a critical point of E on $\mathcal{E}$ if and only if*

$$E(u_t) = E(u) + o(t)$$

for all $\phi \in C_c^\infty(\Omega, \mathbb{R}^N)$.

Our aim is to compute the Euler-Lagrange equation for this kind of varia-
tion. Consider first the derivative $d\Pi_{u(x)}$ at a point $u(x) \in \mathcal{N}$. It is geometrically clear (and can be computed) that $d\Pi_{u(x)}$ is the orthonormal projection
$P(u(x)): \mathbb{R}^N \to T_{u(x)}\mathcal{N}$ onto the tangent space $T_{u(x)}\mathcal{N}$. So we have

$$u_t = u + tP(u).\phi + o(t),$$

and thus

$$
\begin{aligned}
\delta E(u)\left(\frac{du_t}{dt}\bigg|_{t=0}\right) &= \delta E(u)(P(u).\phi) \\
&= \int_\Omega \sum_{\alpha=1}^m \left\langle \frac{\partial u}{\partial x^\alpha}, \frac{\partial}{\partial x^\alpha}(P(u).\phi) \right\rangle dx \\
&= -\int_\Omega \langle \Delta u, P(u).\phi \rangle \, dx \\
&= -\int_\Omega \langle P(u).\Delta u, \phi \rangle \, dx.
\end{aligned}
$$

The Euler-Lagrange equation is therefore

$$P(u).\Delta u = 0,$$

which is just another way to express the condition

$$\Delta u \perp T_u\mathcal{N}.$$

In the case $\mathcal{N} = S^2$, we have seen that harmonic maps satisfy the equation

$$\Delta u + u|du|^2 = 0.$$

A similar condition in the general case would be useful. To derive it, let
$P^\perp(u) = \mathrm{id}_{\mathbb{R}^N} - P(u)$ be the orthonormal projection on $(T_u\mathcal{N})^\perp$. Differen-
tiate the equation

$$P^\perp(u).\frac{\partial u}{\partial x^\alpha} = 0$$

to obtain

$$\sum_{\alpha=1}^m \frac{\partial}{\partial x^\alpha}(P^\perp(u)).\frac{\partial u}{\partial x^\alpha} + P^\perp(u).\Delta u = 0.$$

Hence

$$P^\perp(u).\Delta u = -\sum_{\alpha=1}^m \sum_{j=1}^N \frac{\partial P^\perp(u)}{\partial y^j}.\frac{\partial u^j}{\partial x^\alpha}\frac{\partial u}{\partial x^\alpha} =: -A(u)(du, du).$$

The expression on the right hand side is a quadratic form in du. We call A the second fundamental form of the embedding $\mathcal{N} \subset \mathbb{R}^N$. (As a matter of fact, it coincides with the second fundamental form as already defined for embedded surfaces.) If we apply this equation to a harmonic map, then it yields

$$\Delta u = -A(u)(du, du).$$

So we have indeed a generalization of the S^2-case.

Noether's theorem: The effect of continuous symmetries

Once again our aim is to generalize a fact observed for harmonic maps into S^2. Recall that the harmonic map condition

$$\Delta u \perp T_u S^2$$

for a map $u \colon \Omega \to S^2$ on a domain $\Omega \subset \mathbb{C}$ can be written in the form

$$\frac{\partial}{\partial x}\left(u \times \frac{\partial u}{\partial x}\right) + \frac{\partial}{\partial y}\left(u \times \frac{\partial u}{\partial y}\right) = 0.$$

That is, some divergence term vanishes. Another observation to be made for S^2 is the fact that it is symmetric with regard to rotations. Moreover, this symmetry is compatible with the energy functional in the following way. If $u \colon \Omega \to S^2$ is a map in $\mathcal{E}$ and $R \in \mathrm{SO}(3)$ is given, then Ru is still in $\mathcal{E}$, and

$$E(Ru) = \int_\Omega \frac{|d(Ru)|^2}{2}\, dx = \int_\Omega \frac{|du|^2}{2}\, dx = E(u).$$

Therefore $\mathrm{SO}(3)$ is called a group of symmetries of E on $\{u \colon \Omega \to S^2\}$. We will see that these two observations are connected. Furthermore, this is merely a special case of a general principle, expressed in Noether's theorem.

Let us consider now a variational problem with several variables involving the first derivatives. In the following, Ω is an open domain of $\mathbb{R}^m$ ($m \in \mathbb{N}^*$), we consider the class of maps

$$\mathcal{E} = \{u : \Omega \to \mathbb{R}^n\},$$

and do not pay attention to the regularity of these maps (we could choose for instance these maps in $\mathcal{C}^2$). To any $u \in \mathcal{E}$ we associate the action functional

$$\mathcal{A}[u] = \int_\Omega L(x, u(x), du(x))\, dx.$$

Here, L is a function defined on $\Omega \times \mathbb{R}^n \times M(\mathbb{R}^m, \mathbb{R}^n)$, with values in $\mathbb{R}$, $M(\mathbb{R}^m, \mathbb{R}^n)$ is the set of real matrices $n \times m$, which we identify with the set of linear maps from $\mathbb{R}^m$ to $\mathbb{R}^n$. $du(x)$ is the Jacobian matrix of u, or differential of u at the point x:

$$du = \begin{pmatrix} u_1^1 & \cdots & u_m^1 \\ \vdots & & \vdots \\ u_1^n & \cdots & u_m^n \end{pmatrix} = \begin{pmatrix} \frac{\partial u^1}{\partial x^1} & \cdots & \frac{\partial u^1}{\partial x^m} \\ \vdots & & \vdots \\ \frac{\partial u^n}{\partial x^1} & \cdots & \frac{\partial u^n}{\partial x^m} \end{pmatrix}.$$

Any map u in $\mathcal{E}$ is a critical point of $\mathcal{A}$ if and only if, for any map $v \in \mathcal{E}$ with a compact support in Ω and for $s \in \mathbb{R}$ near 0,

$$\mathcal{A}[u + sv] = \mathcal{A}[u] + o(s).$$

We may then show easily that u is a critical point of $\mathcal{A}$, if and only if u is a solution of the Euler-Lagrange system:

$$\frac{\partial L}{\partial u^i} = \sum_{\alpha=1}^{m} \frac{\partial}{\partial x^\alpha} \left(\frac{\partial L}{\partial u_\alpha^i} \right), \quad \text{for } i = 1, \ldots, n.$$

We want now to see what will happen when such a problem is invariant under the action of a one parameter symmetry group.

Symmetries acting on $\mathbb{R}^n$

It is the most simple case. Assume that there exists a vector field $U : \mathbb{R}^n \to \mathbb{R}^n$ acting on the target space. We denote by Φ_s the flow of U, i.e. the solution of

$$\Phi_0(y) = y, \forall y \in \mathbb{R}^n,$$

$$\frac{d\Phi_s(y)}{ds} = U(\Phi_s(y)), \forall y \in \mathbb{R}^n, \forall s \in \mathbb{R}.$$

The family of diffeomorphisms Φ_s induces a family of deformations of the set of maps from Ω to $\mathbb{R}^n$. These deformations are described geometrically by the action of Φ_s on the graph of a map u: the graph of the deformed map u_s is the image by (id, Φ_s) of the graph of u. Actually it is clear that

$$u_s = \Phi_s \circ u.$$

So we assume the hypothesis that for any sub-domain $\omega \subset \Omega$, and for any map u from ω to $\mathbb{R}^n$,

$$\mathcal{A}_\omega[\Phi_s \circ u] = \mathcal{A}_\omega[u],$$

where $\mathcal{A}_\omega[u] = \int_\omega L(x, u, du)\, dx$. This condition implies in particular that for any ω,

$$\mathcal{A}_\omega[u + sU \circ u] = \mathcal{A}_\omega[u] + o(s).$$

The last relation is true for all ω if and only if the following holds: $\forall (x, y, z) \in \Omega \times \mathbb{R}^n \times M(\mathbb{R}^m, \mathbb{R}^n)$,

$$L(x, y + sU(y), z + sdU(y).z) = L(x, y, z) + o(s). \tag{3.1}$$

Theorem 3.1 *Suppose that U is an infinitesimal symmetry of $\mathcal{A}$, i. e. that (3.1) holds. Let u be a critical point of $\mathcal{A}$, then the vector field J on Ω of components:*

$$J^\alpha(x) = \sum_{i=1}^{n} U^i(u(x)) \frac{\partial L}{\partial u_\alpha^i}(x, u(x), du(x))$$

is divergence free, i. e.

$$\mathrm{div}\, J = \sum_{\alpha=1}^{m} \frac{\partial J^\alpha}{\partial x^\alpha} = 0. \tag{3.2}$$

Remark Let $P_i^\alpha(x, u(x), du(x)) = \frac{\partial L}{\partial u_\alpha^i}(x, u(x), du(x))$ be the impulsion tensor, then J is constructed by contracting P_i^α with $U^i \circ u$, i. e.

$$J^\alpha(x) = \sum_{i=1}^{n} U^i(u(x)) P_i^\alpha(x, u(x), du(x)).$$

Proof. The idea is to use the fact that the action functional $\mathcal{A}(u)$ of u is stationary under the effect of *modulation* of $U(u)$. We choose a function $\phi \in \mathcal{C}_c^1(\Omega, \mathbb{R})$ and we deduce from the critical point condition on u the relation

$$\mathcal{A}[u + s\phi U \circ u] = \mathcal{A}[u] + o(s). \tag{3.3}$$

Let us develop $\mathcal{A}[u + s\phi U \circ u]$. As we shall see it is not necessary to develop the full expression for taking into account the symmetry property of $\mathcal{A}$:

$$
\begin{aligned}
\mathcal{A}[u + s\phi U \circ u] &= \int_\Omega L(x, u + s\phi U(u), du + s\phi dU(u) + sd\phi U(u))\, dx \\
&= \int_\Omega L(x, u + s\phi U(u), du + s\phi dU(u))\, dx \\
&\quad + s \int_\Omega \sum_{i=1}^{n} \sum_{\alpha=1}^{m} U^i(u) \frac{\partial \phi}{\partial x^\alpha} \frac{\partial L}{\partial u_\alpha^i}(x, u, du)\, dx + o(s).
\end{aligned}
$$

Now we use the symmetry relation and in particular the condition (3.1) by substituting $s\phi$ to s. We deduce

$$
\begin{aligned}
\mathcal{A}[u + s\phi U \circ u] &= \mathcal{A}[u] + s \int_\Omega \sum_{i=1}^n \sum_{\alpha=1}^m U^i(u) \frac{\partial \phi}{\partial x^\alpha} \frac{\partial L}{\partial u^i_\alpha}(x, u, du)\, dx + o(s) \\
&= \mathcal{A}[u] + s \int_\Omega \sum_{\alpha=1}^m \frac{\partial \phi}{\partial x^\alpha} J^\alpha\, dx + o(s).
\end{aligned}
$$

If now we compare this last expression with the relation (3.3), we obtain

$$
\int_\Omega \sum_{\alpha=1}^m \frac{\partial \phi}{\partial x^\alpha} J^\alpha\, dx = 0, \forall \phi \in \mathcal{C}_c^1(\Omega, \mathbb{R}).
$$

And this is precisely the weak formulation of our result. $\qquad \square$

Examples

i) Let us first consider the Dirichlet integral in one dimension. Let $\gamma : (a, b) \to \mathbb{R}^n$ be a curve and

$$
\mathcal{A}[\gamma] = \int_a^b m \frac{|\dot\gamma|^2}{2}\, dt.
$$

This corresponds to the problem of finding the trajectory of a free particle of mass m in classical mechanics. Note that $\mathcal{A}$ is invariant by translations in $\mathbb{R}^n$. Since in this case $P_i = m\dot\gamma$ and U^i is a constant, we conclude that the impulsion $m\dot\gamma$ is subject to a conservation law, provided that γ is a critical point of $\mathcal{A}$.

ii) More generally, let L be chosen in a way such that the Legendre condition $\frac{\partial^2 L}{\partial \dot\gamma^2} > 0$ holds. Then we may use the Hamilton formalism. Suppose further that we have a symmetry vector field $U \in \mathcal{C}^1(\mathbb{R}^n, \mathbb{R}^n)$. There are two ways to express this fact:

- $L(t, \gamma + sU(\gamma), \dot\gamma + sdU(\gamma).\dot\gamma) = L(t, \gamma, \dot\gamma) + o(s)$ (like before), or
- $\{\sum_{i=1}^n U^i P^i, H\} = 0$,

where $\{\cdot, \cdot\}$ denotes the Poisson bracket. These two conditions are equivalent. We see that:

- Noether's theorem is immediate in the Hamiltonian setting, for the condition $\{\sum_{i=1}^n U^i P^i, H\} = 0$ implies the conservation law

$$
\frac{d}{dt}\left(\sum_{i=1}^n U^i(\gamma(t)) P^i((\gamma(t)) \right) = 0.
$$

- It is clear that the converse is true.
- We have a generalization of that result: if $F(x, p)$ is chosen such that $\{F, H\} = 0$, then F is conserved.

iii) We return to the energy functional

$$\mathcal{A}[u] = E(u) = \int_\Omega \frac{|du|^2}{2}\, dx$$

for maps $u \colon \Omega \to S^2$. The symmetries of S^2 are given by rotations in $\mathrm{SO}(3)$, which is spanned by the vector fields

$$U_{kl} \colon \mathbb{R}^3 \to \mathbb{R}^3, \quad y \mapsto y^k \frac{\partial}{\partial y^l} - y^l \frac{\partial}{\partial y^k}, \quad 1 \le k < l \le 3.$$

We compute furthermore

$$P_i^\alpha = \frac{\partial u^i}{\partial x^\alpha},$$

and hence

$$J_{kl}^\alpha = u^k \frac{\partial u^l}{\partial x^\alpha} - u^l \frac{\partial u^k}{\partial x^\alpha}$$

are the components of the vector $u \times \frac{\partial u}{\partial x^\alpha}$. Noether's theorem therefore gives

$$\sum_{\alpha=1}^n \frac{\partial}{\partial x^\alpha} \left(u \times \frac{\partial u}{\partial x^\alpha} \right) = 0,$$

a fact that we know already for two-dimensional domains. This shows also how it is related to the symmetry of S^2 via Noether's theorem.

Harmonic maps and conformal transformations

We motivate our next step again first by considering harmonic maps $u \colon \Omega \to \mathcal{N}$, where Ω is a domain in $\mathbb{C}$. Define the complex valued function

$$f = \left| \frac{\partial u}{\partial x} \right|^2 - \left| \frac{\partial u}{\partial y} \right|^2 - 2i \left\langle \frac{\partial u}{\partial x}, \frac{\partial u}{\partial y} \right\rangle.$$

We claim that f is holomorphic. It is convenient to write f in the form

$$f = \left(\frac{\partial u}{\partial x} - i \frac{\partial u}{\partial y} \right)^2 = 4 \left(\frac{\partial u}{\partial z} \right)^2,$$

(where $y^2 = \sum_{i=1}^N (y^i)^2$ for $y \in \mathbb{C}^N$) for this implies immediately

$$\frac{\partial f}{\partial \bar{z}} = 8 \left(\frac{\partial u}{\partial z}, \frac{\partial^2 u}{\partial z \partial \bar{z}} \right) = 2 \left(\frac{\partial u}{\partial z}, \Delta u \right),$$

where $(y_1, y_2) = \sum_{i=1}^{N} y_1^i y_2^i$. The harmonic map condition now implies that $\frac{\partial f}{\partial \bar{z}} = 0$. Note that this fact is equivalent to the system

$$
\begin{cases}
\dfrac{\partial}{\partial x}\left(\left|\dfrac{\partial u}{\partial x}\right|^2 - \left|\dfrac{\partial u}{\partial y}\right|^2 \right) + \dfrac{\partial}{\partial y}\left(2\left\langle \dfrac{\partial u}{\partial x}, \dfrac{\partial u}{\partial y} \right\rangle \right) &= 0, \\[3mm]
\dfrac{\partial}{\partial x}\left(2\left\langle \dfrac{\partial u}{\partial x}, \dfrac{\partial u}{\partial y} \right\rangle \right) - \dfrac{\partial}{\partial y}\left(\left|\dfrac{\partial u}{\partial x}\right|^2 - \left|\dfrac{\partial u}{\partial y}\right|^2 \right) &= 0.
\end{cases}
\tag{3.4}
$$

This means we have two conservation laws. We will see that this is again a consequence of certain symmetries and of (another version of) Noether's theorem.

The symmetries involved in this case are given by conformal maps, for the energy functional is invariant by the action of the conformal group of $\mathbb{C}$ in the following sense. Let $\phi\colon \Omega_2 \to \Omega_1$ be an (oriented) conformal transformation (i. e. $\frac{\partial \phi}{\partial \bar{z}} = 0$ and $\frac{\partial \phi}{\partial z} \neq 0$). Then for any map $u\colon \Omega_1 \to \mathcal{N}$, we have

$$
\int_{\Omega_1} |du|^2 \, dx_1 \, dy_1 = \int_{\Omega_2} |d(u \circ \phi)|^2 \, dx_2 \, dy_2.
$$

The proof of this is based on the two following computations.

- First we compute $|d(u \circ \phi)|^2$:

$$
\begin{aligned}
|d(u \circ \phi)|^2 &= \left|\frac{\partial (u \circ \phi)}{\partial x_2}\right|^2 + \left|\frac{\partial (u \circ \phi)}{\partial y_2}\right|^2 = 4\left(\frac{\partial (u \circ \phi)}{\partial z_2}, \frac{\partial (u \circ \phi)}{\partial \bar{z}_2} \right) \\[2mm]
&= 4\left(\frac{\partial u}{\partial z_1}(\phi), \frac{\partial u}{\partial \bar{z}_1}(\phi) \right) \frac{\partial \phi}{\partial z_2} \frac{\partial \bar{\phi}}{\partial \bar{z}_2} = |du(\phi)|^2 \left|\frac{d\phi}{dz_2}\right|^2,
\end{aligned}
$$

since $\frac{\partial (u \circ \phi)}{\partial z_2} = \frac{\partial u}{\partial z_1}(\phi)\frac{\partial \phi}{\partial z_2} + \frac{\partial u}{\partial \bar{z}_1}(\phi)\frac{\partial \bar{\phi}}{\partial z_2}$ and $\frac{\partial \bar{\phi}}{\partial z_2} = 0$.

- On the other hand, $dx_1 \, dy_1 = \det(d\phi(z_2)) \, dx_2 \, dy_2 = \left|\frac{d\phi}{dz_2}\right|^2 dx_2 \, dy_2$.

Putting these two facts together proves the claim.

We conclude, that a map u is harmonic if and only if its composition with any conformal transformation is. In particular, harmonic maps on Riemannian surfaces depend only on the conformal structure. Eventually, we can apply a version of Noether's theorem (that will follow in the next section) and see a connection of (3.4) with the invariance of the energy by conformal transformations.

Symmetries acting on the domain Ω

As mentioned before, there is a result similar to the version of Noether's theorem that we already know, dealing with the case where $\mathcal{A}$ is invariant by diffeomorphisms acting on Ω. We will have to pay attention however to the fact that the domain might be deformed by such a transformation. Let X be a vector field on an open neighbourhood of Ω, and Ψ_s the flow defined by it. This implies in particular that

$$\Psi_s(x) = x + sX(x) + o(s). \tag{3.5}$$

The image of Ω by Ψ_s is of course in general different from Ω.

Now consider a map $u \colon \Omega \to \mathbb{R}^n$. It is transformed into a map u_s by Ψ_s in a way such that the graph of u_s is the image of the graph of u by (Ψ, id). This means that u_s is defined on $\Omega_s = \Psi_s(\Omega)$, and satisfies

$$u_s \circ \Psi_s = u, \forall s. \tag{3.6}$$

We say that $\mathcal{A}$ is invariant by X if and only if for any subset $\omega \subset \Omega$,

$$\mathcal{A}_{\Psi_s(\omega)}[u_s] = \mathcal{A}_\omega[u].$$

Changing variables $x = \Psi_s(\xi)$ in the left integral, we can write this in the form

$$\int_\omega L\left(\Psi_s(\xi), u_s(\Psi_s(\xi)), du_s(\Psi_s(\xi))\right) \det(d\Psi_s(\xi)) d\xi = \mathcal{A}_\omega[u].$$

On the other hand, a differentiation of (3.6) yields

$$du_s(\Psi_s(\xi)).d\Psi_s(\xi) = du(\xi),$$

respectively

$$du_s(\Psi_s(\xi)) = du(\xi).[d\Psi_s(\xi)]^{-1}.$$

We insert this in the equation above and get

$$\int_\omega L\left(\Psi_s(\xi), u(\xi), du(\xi).[d\Psi_s(\xi)]^{-1}\right) \det(d\Psi_s(\xi)) d\xi = \mathcal{A}_\omega[u]. \tag{3.7}$$

An infinitesimal version of this relation, obtained by developing it up to first order, is then

$$\mathcal{A}_\omega[u] = \int_\omega L\left(x + sX(x), u(x), du(x).[\mathbb{1} - sdX(x)]\right) \det(\mathbb{1} + sdX(x)) dx + o(s).$$

Since this must hold for any ω, we have necessarily $\forall (x, y, z) \in \Omega \times \mathbb{R}^n \times M(\mathbb{R}^m, \mathbb{R}^n)$,

$$L\left(x + sX(x), y, z.[\mathbb{1} - sdX(x)]\right)(1 + sdivX(x)) = L(x, y, z) + o(s). \tag{3.8}$$

Theorem 3.2 *Suppose X is an infinitesimal symmetry of $\mathcal{A}$ in the sense of (3.8). Let u be a critical point of $\mathcal{A}$, then the vector field J on Ω of components*

$$J^\alpha(x) = \sum_{i=1}^n \sum_{\beta=1}^m X^\beta(x) \frac{\partial u^i}{\partial x^\beta} \frac{\partial L}{\partial u^i_\alpha}(x, u(x), du(x)) - X^\alpha(x)L(x, u(x), du(x))$$

is divergence free.

Remark The vector field J^α can be written in the form

$$J^\alpha(x) = \sum_{\beta=1}^m X^\beta(x) H^\alpha_\beta(x),$$

where H^α_β is the Hamiltonian tensor, defined by

$$H^\alpha_\beta(x) = \sum_{i=1}^n \frac{\partial u^i}{\partial x^\beta} \frac{\partial L}{\partial u^i_\alpha}(x, u(x), du(x)) - \delta^\alpha_\beta L(x, u(x), du(x)).$$

Examples

i) Consider the case

$$\mathcal{A}[\gamma] = \int_a^b \left(m\frac{|\dot\gamma|^2}{2} - V(\gamma) \right) dt,$$

where $\gamma\colon (a, b) \to \mathbb{R}^n$ is a curve. This functional is invariant by translation with relation to t. Noether's theorem yields therefore

$$\frac{d}{dt} \left(m\frac{|\dot\gamma|^2}{2} + V(\gamma) \right) = 0,$$

which is known as the energy conservation law in mechanics.

ii) Now consider the two-dimensional case

$$\mathcal{A}[u] = \int_\Omega \frac{|du|^2}{2} \, dx \, dy$$

for a map $u\colon \Omega \to \mathcal{N}$, where $\Omega \subset \mathbb{C}$. Again we have invariance by translations $(x, y) \mapsto (x, y) + \vec{c}$. The Hamiltonian tensor in this situation is

$$H^\alpha_\beta = \sum_{i=1}^N \frac{\partial u^i}{\partial x^\alpha} \frac{\partial u^i}{\partial x^\beta} - \frac{|du|^2}{2} \delta^\alpha_\beta,$$

hence H_α^β are the components of the matrix

$$\frac{1}{2}\left(\begin{array}{cc}\left|\frac{\partial u}{\partial x}\right|^2 - \left|\frac{\partial u}{\partial y}\right|^2 & 2\left\langle\frac{\partial u}{\partial x},\frac{\partial u}{\partial y}\right\rangle \\ 2\left\langle\frac{\partial u}{\partial x},\frac{\partial u}{\partial y}\right\rangle & \left|\frac{\partial u}{\partial y}\right|^2 - \left|\frac{\partial u}{\partial x}\right|^2\end{array}\right).$$

We apply Noether's theorem and obtain

$$\frac{\partial H_\beta^1}{\partial x} + \frac{\partial H_\beta^2}{\partial y} = 0 \quad \text{for } \beta = 1, 2,$$

the two equations corresponding to the vector fields $\frac{\partial}{\partial x}$ and $\frac{\partial}{\partial y}$. But this is again (3.4).

Proof. We build a perturbation of u which is a modulation of the action of X by a function $\phi \in \mathcal{C}_c^1(\Omega, \mathbb{R})$. It gives us a map v_s from Ω to $\mathbb{R}^n$ such that

$$v_s(x + s\phi(x)X(x)) = u(x).$$

Note that if s is sufficiently small, $x \longmapsto x + s\phi(x)X(x)$ is a diffeomorphism of Ω. We may thus perform the change of variables $x = \xi + s\phi(\xi)X(\xi)$ to compute the action of v_s:

$$\begin{aligned}
\mathcal{A}_\Omega[v_s] &= \int_\Omega L(x, v_s(x), dv_s(x))\, dx \\
&= \int_\Omega L\left(\xi + s\phi(\xi)X(\xi), u(\xi), du(\xi).[\mathbb{1} + sd(\phi(\xi)X(\xi))]^{-1}\right) \\
&\qquad\qquad \det[\mathbb{1} + sd(\phi(\xi)X(\xi))]\, d\xi \\
&= \int_\Omega L\left(\xi + s\phi(\xi)X(\xi), u(\xi), du(\xi).[\mathbb{1} - sd(\phi(\xi)X(\xi))]\right) \\
&\qquad\qquad [1 + s\,\mathrm{div}(\phi(\xi)X(\xi))]\, d\xi + o(s).
\end{aligned}$$

Let us develop this expression and then use hypothesis (3.8) (by replacing s by $s\phi(x)$):

$$\begin{aligned}
\mathcal{A}_\Omega[v_s] &= \int_\Omega L\left(x + s\phi(x)X(x), u(x), du(x).[\mathbb{1} - s\phi(x)d(X(x))]\right) \\
&\qquad\qquad [1 + s\phi(x)\mathrm{div}X(x)]\, dx \\
&\quad + s\int_\Omega [-\sum_{\alpha,\beta=1}^m \sum_{i=1}^n \frac{\partial\phi}{\partial x^\alpha}(x)\frac{\partial u^i}{\partial x^\beta}(x)X^\beta(x)\frac{\partial L}{\partial u_\alpha^i}(x, u(x), du(x)) \\
&\qquad\qquad + \sum_{\alpha=1}^m \frac{\partial\phi}{\partial x^\alpha}(x)X^\alpha(x)L(x, u(x), du(x))]\, dx + o(s)
\end{aligned}$$

$$= \; \mathcal{A}_\Omega[u] - s \int_\Omega \sum_{\alpha,\beta=1}^m \frac{\partial \phi}{\partial x^\alpha}(x) X^\beta(x) H^\alpha_\beta(x) \, dx + o(s).$$

Now we just use the fact that u is a critical point of $\mathcal{A}$, and thus that $\mathcal{A}_\Omega[v_s] = \mathcal{A}_\Omega[u] + o(s)$, to deduce

$$-\int_\Omega \sum_{\alpha,\beta=1}^m \frac{\partial \phi}{\partial x^\alpha}(x) X^\beta(x) H^\alpha_\beta(x) \, dx = 0.$$

This shows that J is divergence free. $\qquad \square$

Remark Noether's theorem plays also an important role in proving analytical results on harmonic maps. For more details see [48].

4 Working with the Hopf differential

Change of variables

Let Ω be an open subset of $\mathbb{C}$ and $\mathcal{N} \subset \mathbb{R}^N$ a Riemannian manifold as before. Let $u: \Omega \to \mathcal{N}$ be any map. Define the function

$$f = \left(\frac{\partial u}{\partial x} - i \frac{\partial u}{\partial y} \right)^2 = \left| \frac{\partial u}{\partial x} \right|^2 - \left| \frac{\partial u}{\partial y} \right|^2 - 2i \left\langle \frac{\partial u}{\partial x}, \frac{\partial u}{\partial y} \right\rangle = 4 \left(\frac{\partial u}{\partial z} \right)^2.$$

Recall that f is holomorphic if u is harmonic. Since this property of u is invariant by conformal changes of variables, one might be interested in the behaviour of f by such a transformation. So let's choose a conformal map $\phi: \Omega_2 \to \Omega_1$, where Ω_1 is the domain of a map $u: \Omega_1 \to \mathcal{N}$, and let's check how the corresponding function f transforms. We write z_2 and $z_1 = \phi(z_2)$ for the coordinates of Ω_2 and Ω_1, respectively. Set $f_1(z_1) = 4(\frac{\partial u}{\partial z_1})^2$ and $f_2(z_2) = 4(\frac{\partial (u \circ \phi)}{\partial z_2})^2$. Compute

$$\frac{\partial (u \circ \phi)}{\partial z_2} = \frac{\partial u}{\partial z_1}(\phi) \frac{\partial \phi}{\partial z_2} + \frac{\partial u}{\partial \bar{z}_1}(\phi) \frac{\partial \bar{\phi}}{\partial z_2} = \frac{\partial u}{\partial z_1}(\phi) \frac{\partial \phi}{\partial z_2}.$$

Thus

$$
\begin{aligned}
f_2(z_2) &= 4 \left(\frac{\partial u}{\partial z_1}(\phi(z_2)) \right)^2 \left(\frac{\partial \phi}{\partial z_2}(z_2) \right)^2 \\
&= f_1(\phi(z_2))^2 \left(\frac{\partial \phi}{\partial z_2}(z_2) \right)^2 \\
&= f_1(z_1) \left(\frac{\partial \phi}{\partial z_2}(z_2) \right)^2.
\end{aligned}
$$

We claim that although f is not invariant by ϕ, the so-called Hopf-differential

$$f(z)(dz)^2 = f(z)\, dz \otimes dz \in T^*\Omega \otimes T^*\Omega \otimes \mathbb{C}$$

is. That is,

$$\phi^*(f_1(z_1)(dz_1)^2) = f_2(z_2)(dz_2)^2,$$

where ϕ^* denotes the pull-back by ϕ. By a little abuse of notation we write

$$f_1(z_1)(dz_1)^2 = f_2(z_2)(dz_2)^2.$$

This property of the Hopf differential can easily be checked. We have

$$
\begin{aligned}
\phi^*(dz_1)^2 &= (d(\phi(z_2)))^2 = \left(\frac{\partial \phi}{\partial z_2} \right)^2 (dz_2)^2, \\
\phi^*(f_1(z_1)) &= (f_1 \circ \phi)(z_2).
\end{aligned}
$$

Together with the earlier computation this proves indeed the claim.

Applications

1) Harmonic maps from the sphere S^2

Recall that there exists a map, called the stereographic projection, that maps the sphere S^2 except one point conformaly to $\mathbb{C}$. So if we write $S^2 = U_1 \cup U_2$, where

$$U_1 = S^2 \backslash \{(0,0,-1)\}, \quad U_2 = S^2 \backslash \{(0,0,1)\},$$

and identify both U_1 and U_2 with $\mathbb{C}$ via the corresponding stereographic projections ψ_1 and ψ_2, then this gives a conformal structure on S^2. Therefore we say that a map $u \colon S^2 \to \mathcal{N}$ is harmonic if $u \circ \psi_1^{-1}$ and $u \circ \psi_2^{-1}$ are so.

The stereographic projections ψ_1 and ψ_2 are given by

$$\psi_1^{-1}(z) = \frac{1}{1+|z|^2} \begin{pmatrix} z + \bar{z} \\ -i(z - \bar{z}) \\ 1 - |z|^2 \end{pmatrix}, \quad \psi_2^{-1} = \frac{1}{1+|z|^2} \begin{pmatrix} z + \bar{z} \\ i(z - \bar{z}) \\ -1 + |z|^2 \end{pmatrix}.$$

It is easy to check that $\psi_2^{-1}(1/z) = \psi_1^{-1}(z)$ for $z \in \mathbb{C}^*$. Consider therefore the conformal map

$$\phi \colon \mathbb{C}^* \to \mathbb{C}^*, \quad z \mapsto \frac{1}{z}$$

and the two harmonic maps $v_1 = u \circ \psi_1^{-1}$ and $v_2 = u \circ \psi_2^{-1}$, which satisfy $v_2 = v_1 \circ \phi$ on $\mathbb{C}^*$. Let

$$f_1(z_1) = 4 \left(\frac{\partial v_1}{\partial z_1} \right)^2 (z_1), \quad f_2(z_2) = 4 \left(\frac{\partial v_2}{\partial z_2} \right)^2 (z_2).$$

Then the Hopf differentials satisfy

$$f_1(z_1)(z_1)^2 = f_2(z_2)(dz_2)^2 = f_2(1/z_1) \frac{(dz_1)^2}{z_1^4}.$$

This means

$$f_1(z_1) = f_2(1/z_1) \frac{1}{z_1^4}$$

for all $z_1 \in \mathbb{C}^*$. As $z_1 \to \infty$, we get

$$\lim_{z_1 \to \infty} f_1(z_1) = 0.$$

But recall that f_1 is holomorphic. Liouville's theorem therefore implies that $f \equiv 0$. Thus u is weakly conformal.

Theorem 4.1 *Any harmonic map $u \colon S^2 \to \mathcal{N}$ is weakly conformal.*

Using this we can prove the following.

Corollary 4.1 *Any harmonic map* $u: S^2 \to S^2$ *is either*

    i) constant, or

  ii) holomorphic, or

 iii) antiholomorphic.

The proof of this Corollary uses the following lemma. We prove both results in the appendix, at the end of the chapter. (The lemma is more or less well-known, another proof can be found in [45].)

Lemma 4.1 *Let* $U, V \subset \mathbb{C}$ *be two open subsets of* $\mathbb{C}$ *and* $u: U \to V$ *a weakly conformal function of class* C^2. *If* U *is connected, then* u *is either a constant, or holomorphic, or antiholomorphic.*

2) Hopf's theorem

Theorem 4.2 *Any CMC immersion of the sphere* S^2 *in* $\mathbb{R}^3$ *is a round sphere.*

This was proved by H. Hopf in 1951 [55]. A. D. Aleksandrov [5] proved in 1956, that any embedded compact CMC surface in $\mathbb{R}^3$ without boundary is a round sphere. This led to the so-believed Hopf-conjecture (which was not formulated by Hopf) that there exist no immersed compact CMC surfaces besides the sphere at all in $\mathbb{R}^3$. However, H. Wente [86] found in 1984 an immersed CMC torus, and later N. Kapouleas [56] constructed in 1987 immersed CMC surfaces of even higher genus.

Proof of Hopf's theorem. Let $X: S^2 \to \mathbb{R}^3$ be a conformal CMC immersion, and let $u: S^2 \to S^2$ be its Gauss map. In local conformal coordinates, we may write for the fundamental forms of X

$$I = e^{2\omega} \begin{pmatrix} 1 & 0 \\ 0 & 1 \end{pmatrix}, \quad II = e^{2\omega} \begin{pmatrix} H+a & b \\ b & H-a \end{pmatrix} = -{}^t du.dX,$$

as before. Recall that we have

$$\begin{cases} \dfrac{\partial u}{\partial x} &= -(H+a)\dfrac{\partial X}{\partial x} - b\dfrac{\partial X}{\partial y}, \\[2mm] \dfrac{\partial u}{\partial y} &= -b\dfrac{\partial X}{\partial x} - (H-a)\dfrac{\partial X}{\partial y}, \end{cases}$$

or, equivalently,

$$\frac{\partial u}{\partial z} = -H\frac{\partial X}{\partial z} - (a - ib)\frac{\partial X}{\partial \bar{z}}. \tag{4.1}$$

Now compute

$$f = 4\left(\frac{\partial u}{\partial z}\right)^2$$

$$= 4\left[H^2\left(\frac{\partial X}{\partial z}\right)^2 + (a-ib)^2\left(\frac{\partial X}{\partial \bar{z}}\right)^2 + 2H(a-ib)\left(\frac{\partial X}{\partial z}, \frac{\partial X}{\partial \bar{z}}\right)\right].$$

Since X is conformal, the expressions $\left(\frac{\partial X}{\partial z}\right)^2$ and $\left(\frac{\partial X}{\partial \bar{z}}\right)^2$ both vanish. Furthermore,

$$\left(\frac{\partial X}{\partial z}, \frac{\partial X}{\partial \bar{z}}\right) = \left|\frac{\partial X}{\partial z}\right|^2 = \frac{e^{2\omega}}{2}.$$

So we have

$$f = 4H(a-ib)e^{2\omega}.$$

But we know that u is harmonic, for X is a CMC immersion. Hence by Theorem 4.1

$$H(a-ib) \equiv 0,$$

which implies either

- $H = 0$ and thus $\Delta X \equiv 0$, or
- $a \equiv b \equiv 0$, i. e. the surface is completely umbilic. In this case we have

$$\mathrm{II} = e^{2\omega}H\begin{pmatrix} 1 & 0 \\ 0 & 1 \end{pmatrix}.$$

In the first case we apply the maximum principle to find that X is constant and hence no immersion. Thus this doesn't occur. In the second case we have

$$\frac{\partial}{\partial z}(u + HX) = 0$$

because of (4.1). But $u + HX$ is real. So $d(u + HX) = 0$, hence $u + HX$ is constant, say

$$u + HX \equiv C \in \mathbb{R}^3.$$

This means

$$\left|X - \frac{C}{H}\right|^2 = \frac{|u|^2}{H^2} = \frac{1}{H^2},$$

which proves that the surface parametrized by X is indeed a subset of a round sphere (of center C/H and radius $1/H$). Lastly we can apply Corollary 4.1 to X and we conclude using the fact that X is an immersion. $\square$

Remark We obtain the formula

$$f = 4H(a - ib)e^{2\omega}$$

for any conformal CMC immersion $X : \Omega \to \mathbb{R}^3$, where Ω is an open subset of $\mathbb{C}$. Suppose $H \neq 0$. Then we have either

- $f \equiv 0$, i. e. $a \equiv b \equiv 0$, in which case the surface is totally umbilic and hence a sphere, or
- $f \not\equiv 0$. Then f has (at most) isolated zeros. Hence the umbilic points are isolated.

Non-umbilic case

Assume that $X : \Omega \to \mathbb{R}^3$ is a conformal CMC immersion without umbilics. Let

$$f(z) = 4 \left(\frac{\partial u}{\partial z} \right)^2 (z)$$

be the component along $(dz)^2$ of the Hopf differential of u. By hypothesis, f has no zeros. Assume $H = \frac{1}{2}$. Let $\psi : \Omega \to \mathbb{C}$ be a holomorphic solution of

$$\frac{d\psi}{dz}(z) = \sqrt{-f(z)}.$$

We may consider $Y = X \circ \psi^{-1}$ instead of X, at least locally, for this is just another conformal parametrization of the surface. Its Gauss map is $v = u \circ \psi^{-1}$. Now,

$$
\begin{aligned}
4 \left(\frac{\partial v}{\partial \zeta}(\zeta) \right)^2 &= 4 \left(\frac{\partial u}{\partial z}(\psi^{-1}(\zeta)) \right)^2 \left(\frac{d\psi^{-1}}{d\zeta}(\zeta) \right)^2 \\
&= 4 \left(\frac{\partial u}{\partial z}(z) \right)^2 \left(\frac{d\psi}{dz}(z) \right)^{-2} \\
&= f(z)(-f(z))^{-1} = -1.
\end{aligned}
$$

Let us work now with these variables. We then obtain

$$2(a - ib)e^{2\omega} = 4H(a - ib)e^{2\omega} = -1,$$

which means $a = -\frac{e^{-2\omega}}{2}$ and $b = 0$. We may thus write

$$\mathrm{II} = e^{2\omega} \begin{pmatrix} H + a & b \\ b & H - a \end{pmatrix} = e^{\omega} \begin{pmatrix} \sinh \omega & 0 \\ 0 & \cosh \omega \end{pmatrix}.$$

Conclusion. The x-lines (images by Y of the straight lines $y = $ const) are curvature lines for the smaller principal curvature, the y-lines are curvature lines for the larger principal curvature.

4.1 Appendix

Here, we give the proofs of corollary 4.1 and lemma 4.1.

Lemma 4.1 *Let U be an open connected domain in $\mathbb{C}$ and let $u : U \longrightarrow \mathbb{C}$ be a map of class $\mathcal{C}^2$. Assume that u is weakly conformal, i. e.*

$$|\partial_x u|^2 - |\partial_y u|^2 - 2i\langle \partial_x u, \partial_y u \rangle = 0. \tag{4.2}$$

Then, either u is constant, or holomorphic, or antiholomorphic.

Proof. We decompose $U = \mathcal{C} \sqcup U^- \sqcup U^+$, where

$$\mathcal{C} := \{z \in U : \ du_z = 0\},$$

and $U^- = \{z \in U : \ \det du_z < 0\}$, $U^+ = \{z \in U : \ \det du_z > 0\}$. Since u is $\mathcal{C}^2$, $\mathcal{C}$ is clearly closed and U^- and U^+ are open. The fact that this is a partition of U follows from the fact that, by (4.2), for any $z \in U$, du_z is either of rank 0, or of rank 2. Moreover, simple algebra using (4.2) implies that u is holomorphic in U^+ and is antiholomorphic in U^-.

We exclude the situation where $\mathcal{C} = U$, which corresponds to the case where u is constant. Our task will be then to prove that $\mathcal{C}$ is composed of isolated points. Then, $U^- \sqcup U^+ = U \setminus \mathcal{C}$ is connected and thus $\det du$ has a constant sign on it, which means that either $U \setminus \mathcal{C} = U^-$, or $U \setminus \mathcal{C} = U^+$ and the conclusion follows easily.

Denote $u(z) = u^1(z) + iu^2(z)$ and derivate (4.2) with respect to $\bar{z}$. We obtain

$$\begin{aligned} 0 &= \left(\frac{\partial}{\partial x} + i \frac{\partial}{\partial y} \right) \sum_{j=1}^{2} \left(\frac{\partial u^j}{\partial x} - i \frac{\partial u^j}{\partial y} \right)^2 \\ &= 2 \sum_{j=1}^{2} \Delta u^j \left(\frac{\partial u^j}{\partial x} - i \frac{\partial u^j}{\partial y} \right). \end{aligned}$$

Thus we deduce that $\Delta u = 0$ on $U \setminus \mathcal{C}$.

Now let $\alpha := -\frac{\partial u^1}{\partial y} dx + \frac{\partial u^1}{\partial x} dy$. We remark that $d\alpha = \Delta u^1 dx \wedge dy$, hence

$$d\alpha = 0 \text{ on } U \setminus \mathcal{C}. \tag{4.3}$$

But we also have

$$\mathcal{C} = \{z \in U : \ \alpha_z = 0\}. \tag{4.4}$$

We claim that actually $d\alpha$ vanishes at any point $z_0 \in U$.

- suppose that z_0 is interior to $\mathcal{C}$. Then by (4.4) there exists a ball $B(z_0, r) \subset \mathcal{C}$, on which $\alpha = 0$ and thus $d\alpha = 0$ on $B(z_0, r)$.

- suppose that z_0 is not interior to C. Then there exists a sequence $(\zeta_n)_{n \in \mathbb{N}}$ of points in $U \setminus C$ which converges to z_0. By (4.3), $d\alpha_{\zeta_n} = 0$, and since u is C^2, $d\alpha$ is continuous and thus $d\alpha_{z_0} = 0$.

Hence α is closed on U. Thus we can locally integrate $\alpha = dg$, for some real valued function g. It turns out that $u^1 + ig$ is a non-constant holomorphic function of z and thus its critical set, C, is composed of isolated points.

Corollary 4.1 *Let $u : S^2 \longrightarrow S^2$ be a weakly conformal map of class C^2. Then*

- *either u is constant,*
- *or $u = Q_s \circ v \circ P_m$ or $u = Q_s \circ \bar{v} \circ P_m$, where $P_m : S^2 \longrightarrow \mathbb{C} \cup \{\infty\}$ and $Q_s : \mathbb{C} \cup \{\infty\} \longrightarrow S^2$ are stereographic and inverse stereographic projections respectively and $v : \mathbb{C} \cup \{\infty\} \longrightarrow \mathbb{C} \cup \{\infty\}$ is a (meromorphic) rational function of z.*

Proof. Using Sard's theorem, we can select a point $s \in S^2$ which is a regular value of u, then we let $P_s : S^2 \longrightarrow \mathbb{C} \cup \{\infty\}$ be a stereographic projection mapping s to ∞. Similarly we choose some point m such that $u(m) \neq s$ and we let $Q_m : \mathbb{C} \cup \{\infty\} \longrightarrow S^2$ be an inverse stereographic projection such that $Q_m(\infty) = m$. Then $v := P_s \circ u \circ Q_m$ is weakly conformal from $\mathbb{C} \setminus \{a_1, \ldots, a_p\}$ to $\mathbb{C}$, where $\{a_1, \ldots, a_p\}$ is the counter-image of s by $u \circ Q_m$. Lemma 4.1 above proves that v is either constant, or holomorphic, or antiholomorphic. We exclude the constant case and without loss of generality, we shall assume that v is holomorphic. Then, since s is a regular value of u, each point a_k is a simple pole of v and v is a meromorphic function on $\mathbb{C}$ which tends to a constant $P_s(u(m))$ at infinity.

Let $\{b_1, \ldots, b_q\}$ be the zeroes of v and denote $d_l \in \mathbb{N}^\star$ the multiplicity of each b_l. Then, choosing a radius $R > 0$ such that $a_1, \ldots, a_p, b_1, \ldots, b_q$ are contained in $B(0, R)$ and computing $0 = \int_{\partial B(0,R)} \frac{df}{f}$ using residue formula, we show that $\sum_{l=1}^{q} d_l = \sum_{k=1}^{p} 1 = p$. Lastly we apply Liouville's theorem to the function

$$z \longmapsto f(z) = \frac{(z - a_1) \ldots (z - a_p)}{(z - b_1)^{d_1} \ldots (z - b_q)^{d_q}} v(z),$$

to deduce that $f = c = P_s(u(m))$ everywhere. This terminates the proof with $Q_s = P_s^{-1}$, $P_m = Q_m^{-1}$ and $v(z) = c \frac{(z-b_1)^{d_1} \ldots (z-b_q)^{d_q}}{(z-a_1) \ldots (z-a_p)}$.

5 The Gauss-Codazzi condition

We have seen at the end of the previous chapter, that given any conformal CMC immersion $X: \Omega \to \mathbb{R}^3$ without umbilic points, with Gauss map $u: \Omega \to S^2$, and with $H \neq 0$, we can find a new parametrization $Y = 2HX \circ \psi^{-1}$, such that

$$\mathrm{I} = e^{2\omega} \begin{pmatrix} 1 & 0 \\ 0 & 1 \end{pmatrix}, \quad \mathrm{II} = e^{\omega} \begin{pmatrix} \sinh \omega & 0 \\ 0 & \cosh \omega \end{pmatrix}.$$

Moreover, as we will see later, ω satisfies the compatibility condition

$$\Delta \omega + \sinh \omega \cosh \omega = 0. \tag{5.1}$$

There is a converse result to that. Any solution of (5.1) on a simply connected domain gives rise to a "normalized" CMC immersion. We will prove that later, by considering a more general kind of inverse problem as follows. Given

$$\mathrm{I} = e^{2\omega} \begin{pmatrix} 1 & 0 \\ 0 & 1 \end{pmatrix} \quad \text{and} \quad \mathrm{II} = e^{\omega} \begin{pmatrix} H+a & b \\ b & H-a \end{pmatrix},$$

where H is a constant and $a, b, \omega: \Omega \to \mathbb{R}$ are certain functions, is there a CMC immersion that has I and II as its fundamental forms?

Gauss-Codazzi condition as a zero-curvature condition

We want to find necessary and sufficient conditions such that the question above has an affirmative answer. Suppose first that X is a CMC immersion corresponding to I and II as given above. Introduce the moving frame $(e_1, e_2, e_3): \Omega \to \mathrm{SO}(3, \mathbb{R})$, where $dX = e^{\omega}(e_1 dx + e_2 dy)$, and $u = e_3 = e_1 \times e_2$ is the Gauss map of X. Write

$$de_i = \sum_{j=1}^{3} A_i^j e_j, \quad i = 1, 2, 3,$$

$$dX = \sum_{j=1}^{3} A^j e_j.$$

That is, the 1-forms A_i^j and A^j are the components of de_i and dX, respectively, with respect to the basis (e_1, e_2, e_3) of $\mathbb{R}^3$.

Remarks

i) We know already that $dX = e^{\omega}(e_1 dx + e_2 dy)$, so

$$\begin{pmatrix} A^1 \\ A^2 \\ A^3 \end{pmatrix} = e^{\omega} \begin{pmatrix} dx \\ dy \\ 0 \end{pmatrix}.$$

ii) The fact that $\langle e_a, e_b \rangle = \delta_{ab}$ implies that the matrix with components A_a^b is skew symmetric, i. e. $A_a^b + A_b^a = 0$ for $1 \leq a, b \leq 3$.

iii) Consider the (4×4)-matrix

$$F = \begin{pmatrix} e_1 & e_2 & e_3 & X \\ 0 & 0 & 0 & 1 \end{pmatrix}.$$

Note that the part $R = (e_1, e_2, e_3)$ of this matrix is in $SO(3, \mathbb{R})$. Define furthermore the matrix

$$A = \begin{pmatrix} 0 & A_2^1 & A_3^1 & A^1 \\ A_1^2 & 0 & A_3^2 & A^2 \\ A_1^3 & A_2^3 & 0 & A^3 \\ 0 & 0 & 0 & 0 \end{pmatrix},$$

that has 1-forms as component. Then we can compute

$$dF = FA. \tag{5.2}$$

We may think of F as an element of $SO(3, \mathbb{R}) \ltimes \mathbb{R}^3$, and analogously of A as being in the corresponding Lie algebra $so(3, \mathbb{R}) \oplus \mathbb{R}^3$. Note that $SO(3, \mathbb{R}) \ltimes \mathbb{R}^3$ acts on $\mathbb{R}^3$ in the following way. For any $\xi \in \mathbb{R}^3$,

$$F \begin{pmatrix} \xi \\ 1 \end{pmatrix} = \begin{pmatrix} R\xi + X \\ 1 \end{pmatrix}.$$

Thus $SO(3, \mathbb{R}) \ltimes \mathbb{R}^3$ may be seen as the group of affine isometries of $\mathbb{R}^3$.

Lemma 5.1 *Assume that X is a CMC immersion such that*

$$I = e^{2\omega} \begin{pmatrix} 1 & 0 \\ 0 & 1 \end{pmatrix}, \quad II = e^{2\omega} \begin{pmatrix} H + a & b \\ b & H - a \end{pmatrix},$$

and

$$f := 4 \left(\frac{\partial u}{\partial z} \right)^2 = 4H(a - ib)e^{2\omega}.$$

Then

$$A = \begin{pmatrix} 0 & -*d\omega & -A_1^3 & e^\omega \, dx \\ *d\omega & 0 & -A_2^3 & e^\omega \, dy \\ A_1^3 & A_2^3 & 0 & 0 \\ 0 & 0 & 0 & 0 \end{pmatrix}, \tag{5.3}$$

where

$$A_1^3 = e^\omega H \, dx + \frac{e^{-\omega}}{4H} \mathrm{Re}(f \, dz), \quad A_2^3 = e^\omega H \, dy + \frac{e^{-\omega}}{4H} \mathrm{Re}(if \, dz), \tag{5.4}$$

and $*: T^*\mathbb{R}^2 \to T^*\mathbb{R}^2$ denotes the Hodge operator. In particular $*d\omega = -\frac{\partial \omega}{\partial y} \, dx + \frac{\partial \omega}{\partial x} \, dy$.

Proof. Only the components A_1^2, A_1^3, and A_2^3 need to be computed, the rest then follows from the definition of A. Since this can be done quite analogously for A_1^3 and A_2^3, we will confine ourselves to computing one of them. We have

$$A_1^2 = \left\langle \frac{\partial e_1}{\partial x}, e_2 \right\rangle dx + \left\langle \frac{\partial e_1}{\partial y}, e_2 \right\rangle dy = -\left\langle \frac{\partial e_2}{\partial x}, e_1 \right\rangle dx + \left\langle \frac{\partial e_1}{\partial y}, e_2 \right\rangle dy.$$

But since

$$0 = d^2 X = d(e^\omega (e_1\, dx + e_2\, dy)) = e^\omega \left(e_2 \frac{\partial \omega}{\partial x} + \frac{\partial e_2}{\partial x} - e_1 \frac{\partial \omega}{\partial y} - \frac{\partial e_1}{\partial y} \right) dx \wedge dy,$$

we see that

$$\left\langle e_1, \frac{\partial e_2}{\partial x} \right\rangle = \frac{\partial \omega}{\partial y},$$
$$\left\langle e_2, \frac{\partial e_1}{\partial y} \right\rangle = \frac{\partial \omega}{\partial x}.$$

Thus indeed

$$A_1^2 = -\frac{\partial \omega}{\partial y}\, dx + \frac{\partial \omega}{\partial x}\, dy = *d\omega.$$

Now compute

$$A_1^3 = \langle de_1, e_3 \rangle = -\langle du, e_1 \rangle = -e^{-\omega} \left\langle du, \frac{\partial X}{\partial x} \right\rangle = e^\omega ((H + a)\, dx + b\, dy)$$

(recall (2.2)). This can be expressed in terms of f by

$$A_1^3 = e^\omega H\, dx + \frac{e^{-\omega}}{4H} \mathrm{Re}(f\, dz),$$

which concludes the proof. □

Now, take the exterior derivative of (5.2). This yields

$$0 = d^2 F = d(FA) = F dA + dF \wedge A = F(dA + A \wedge A).$$

(See Section 7.1 for more details on this notation.) Since F is clearly invertible, we obtain the necessary condition

$$dA + A \wedge A = 0 \tag{5.5}$$

for X to be a CMC immersion.

Remark Think of A as a connection form, then $dA + A \wedge A$ is the corresponding curvature 2-form. Thus (5.5) is a zero curvature condition.

Converse result

We have obtained (5.5) as a consequence of (5.2). The following lemma now states a kind of converse result. The proof will be given in the appendix of this chapter.

Lemma 5.2 *Let Ω be a simply connected open subset of $\mathbb{R}^m$ and $x_0 \in \Omega$. Let $A \in \mathcal{C}^1(\Omega, T^*\mathbb{R}^m \otimes \mathrm{gl}(n, \mathbb{R}))$, where $\mathrm{gl}(n, \mathbb{R})$ is the Lie algebra of the linear Lie group $\mathrm{GL}(n, \mathbb{R})$ (i. e. $\mathrm{gl}(n, \mathbb{R})$ consists of all $(n \times n)$-matrices with real components). Write $A = \sum_{\alpha=1}^m A_\alpha \, dx^\alpha$. Assume that*

$$dA + A \wedge A = \sum_{\alpha < \beta} \left(\frac{\partial A_\beta}{\partial x^\alpha} - \frac{\partial A_\alpha}{\partial x^\beta} + [A_\alpha, A_\beta] \right) dx^\alpha \wedge dx^\beta = 0.$$

Then for any $F_0 \in \mathrm{GL}(n, \mathbb{R})$ there exists a unique $F \in \mathcal{C}^2(\Omega, \mathrm{GL}(n, \mathbb{R}))$ such that

$$\begin{cases} F(x_0) &= F_0, \\ dF &= FA. \end{cases}$$

Remark The Lie group $\mathrm{GL}(n, \mathbb{R})$ and its Lie algebra $\mathrm{gl}(n, \mathbb{R})$ could be replaced by some other pair of Lie group and Lie algebra, and the lemma would still be true, e. g. respectively by $\mathrm{SL}(n, \mathbb{R})$ and $\mathrm{sl}(n, \mathbb{R})$, $\mathrm{SO}(n, \mathbb{R})$ and $\mathrm{so}(n, \mathbb{R})$, or in $\mathrm{GL}(n, \mathbb{C})$, $\mathrm{U}(n)$ and $\mathrm{u}(n)$. In particular, if we have

$$A = \begin{pmatrix} 0 & -A_1^2 & -A_1^3 & A^1 \\ A_1^2 & 0 & -A_2^3 & A^2 \\ A_1^3 & A_2^3 & 0 & A^3 \\ 0 & 0 & 0 & 0 \end{pmatrix} \in \mathrm{so}(3, \mathbb{R}) \oplus \mathbb{R}^3$$

and

$$F_0 = \begin{pmatrix} e_{1,0} & e_{2,0} & e_{3,0} & X_0 \\ 0 & 0 & 0 & 1 \end{pmatrix} \in \mathrm{SO}(3, \mathbb{R}) \ltimes \mathbb{R}^3,$$

then the F obtained by lemma 5.2 takes values in $\mathrm{SO}(3, \mathbb{R}) \ltimes \mathbb{R}^3$ everywhere.

Application

Lemma 5.3 *Suppose that A has the form (5.3), (5.4), where H is a constant, $\omega: \Omega \to \mathbb{R}$ some function, and f holomorphic on Ω. Then*

$$dA + A \wedge A = \left(\Delta\omega + e^{2\omega}H^2 - \frac{e^{-2\omega}|f|^2}{(4H)^2} \right) \begin{pmatrix} 0 & -1 & 0 & 0 \\ 1 & 0 & 0 & 0 \\ 0 & 0 & 0 & 0 \\ 0 & 0 & 0 & 0 \end{pmatrix} dx \wedge dy.$$

This can be proved by a mere computation, using the fact that $d(\text{Re}(f\,dz)) = d(\text{Re}(if\,dz)) = 0$, since f is holomorphic.

Now we put all facts together. On the one hand, if X is a CMC immersion, then (5.5) and lemma 5.3 show that

$$\Delta\omega + e^{2\omega}H^2 - \frac{e^{-2\omega}|f|^2}{(4H)^2} = 0. \tag{5.6}$$

On the other hand, suppose that (5.6) is true for a constant H, a function $\omega : \Omega \to \mathbb{R}$ and a holomorphic function f. Then if A has the form (5.3), (5.4), it satisfies $dA + A \wedge A = 0$. Furthermore, it takes values in $\text{so}(3,\mathbb{R}) \oplus \mathbb{R}^3$. We can therefore find

$$F = \begin{pmatrix} e_1 & e_2 & e_3 & X \\ 0 & 0 & 0 & 1 \end{pmatrix}$$

with values in $\text{SO}(3,\mathbb{R}) \ltimes \mathbb{R}^3$, such that $dF = FA$, according to lemma 5.2. It is now easy to see that X is a CMC immersion with mean curvature H. Thus (5.6) is a necessary and sufficient condition for ω, H, and f to belong to a CMC immersion.

Note that in the non-umbilic case and in the "normalized" situation $H = \frac{1}{2}$, $|f| = 1$ (we know that we can even achieve $f = -1$), (5.6) takes the form

$$\Delta\omega + \sinh\omega\cosh\omega = 0.$$

The corresponding second fundamental form is then

$$\text{II} = e^{2\omega}\begin{pmatrix} \sinh\omega & 0 \\ 0 & \cosh\omega \end{pmatrix}.$$

Conjugate minimal surfaces and conjugate CMC surfaces

Recall Weierstrass representation

$$X(z) = \text{Re}\left(\int_{z_0}^{z} h \begin{pmatrix} \frac{i}{2}(v^2 - 1) \\ \frac{1}{2}(v^2 + 1) \\ iv \end{pmatrix}(\zeta)\,d\zeta\right)$$

for a minimal surface. We have seen in Chapter 2 that if u is the Gauss map of X and P a stereographic projection, then we may choose $v = P \circ (-u)$ and

$$h = \frac{4i}{(a+ib)(1+|v|^2)}\frac{\partial\bar{v}}{\partial\bar{z}}.$$

For any $\lambda \in S^1 \subset \mathbb{C}^*$, we can modify this formula as follows. Set

$$X_\lambda(z) = \text{Re}\left(\int_{z_0}^z \lambda^{-2} h \begin{pmatrix} \frac{i}{2}(v^2 - 1) \\ \frac{1}{2}(v^2 + 1) \\ iv \end{pmatrix} (\zeta)\, d\zeta \right).$$

This is equivalent to replacing $a - ib$ by $\lambda^{-2}(a - ib)$. We get thus a 1-parameter family of minimal surfaces, all with Gauss map u, called the conjugate family.

Now suppose that X is a CMC immersion. There is no simple formula similar to Weierstrass representation for this case. However, the following may be done instead. Replacing $(a - ib)$ by $\lambda^{-2}(a - ib)$ means replacing f by $\lambda^{-2}f$. Keeping H and ω fixed, this gives rise to a deformation A_λ of A. But...

Key observation: we still have

$$dA_\lambda + A_\lambda \wedge A_\lambda = 0,$$

according to lemma 5.3, since we have not changed $|f|$. Hence also in this case we can find a conjugate family of CMC immersion, coming from this A_λ. However, the Gauss map needn't stay the same. This is a result due to O. Bonnet [18].

We are going to look at this construction a little more closely. Write

$$A_\lambda = \begin{pmatrix} \breve{A}_\lambda & \phi_\lambda \\ 0 & 0 \end{pmatrix},$$

where $\breve{A}_\lambda$ is a (3×3)-matrix and ϕ_λ a column vector. Since

$$\phi_\lambda = e^\omega \begin{pmatrix} dx \\ dy \\ 0 \end{pmatrix},$$

this part of A_λ doesn't bear much information. The important part will therefore be $\breve{A}_\lambda$. We compute

$$dA_\lambda + A_\lambda \wedge A_\lambda = \begin{pmatrix} d\breve{A}_\lambda + \breve{A}_\lambda \wedge \breve{A}_\lambda & d\phi_\lambda + \breve{A}_\lambda \wedge \phi_\lambda \\ 0 & 0 \end{pmatrix}.$$

The zero curvature condition

$$dA_\lambda + A_\lambda \wedge A_\lambda = 0$$

for A_λ implies therefore

$$d\breve{A}_\lambda + \breve{A}_\lambda \wedge \breve{A}_\lambda = 0.$$

Furthermore, from lemma 5.1 and from the construction of A_λ, we see that

$$\check{A}_\lambda = \begin{pmatrix} 0 & -*d\omega & -\left(\frac{e^\omega}{2}dx + \frac{e^{-\omega}}{2}\mathrm{Re}(\lambda^{-2}f\,dz)\right) \\ *d\omega & 0 & -\left(\frac{e^\omega}{2}dy + \frac{e^{-\omega}}{2}\mathrm{Re}(i\lambda^{-2}f\,dz)\right) \\ \frac{e^\omega}{2}dx + \frac{e^{-\omega}}{2}\mathrm{Re}(\lambda^{-2}f\,dz) & \frac{e^\omega}{2}dy + \frac{e^{-\omega}}{2}\mathrm{Re}(i\lambda^{-2}f\,dz) & 0 \end{pmatrix},$$

if we assume $H = \frac{1}{2}$. Finally, (5.2) corresponds to

$$d\check{F}_\lambda = \check{F}_\lambda \check{A}_\lambda, \tag{5.7}$$

where

$$\check{F}_\lambda := (e_{1,\lambda}, e_{2,\lambda}, e_{3,\lambda})$$

and where $e_{1,\lambda}, e_{2,\lambda}, e_{3,\lambda}$ are the deformations of e_1, e_2, e_3 which arise with the replacing of f by $\lambda^{-2}f$ as described above.

We want to modify (5.7) a little further. Choose $\theta \in \mathbb{R}$ such that $\lambda = e^{i\theta}$. Consider the rotation

$$R_\lambda = \begin{pmatrix} \cos\theta & -\sin\theta & 0 \\ \sin\theta & \cos\theta & 0 \\ 0 & 0 & 1 \end{pmatrix} = \begin{pmatrix} \frac{\lambda+\lambda^{-1}}{2} & \frac{\lambda^{-1}-\lambda}{2i} & 0 \\ \frac{\lambda-\lambda^{-1}}{2i} & \frac{\lambda+\lambda^{-1}}{2} & 0 \\ 0 & 0 & 1 \end{pmatrix},$$

and set

$$G_\lambda := \check{F}_\lambda R_\lambda,$$

and

$$\alpha_\lambda := R_\lambda^{-1}\check{A}_\lambda R_\lambda.$$

Then (5.7) transforms into

$$dG_\lambda = G_\lambda \alpha_\lambda.$$

Moreover, the following can be computed:

$$\alpha_\lambda = \frac{\lambda^{-1}}{4}\begin{pmatrix} 0 & 0 & -e^\omega - e^{-\omega}f \\ 0 & 0 & ie^\omega - ie^{-\omega}f \\ e^\omega + e^{-\omega}f & -ie^\omega + ie^{-\omega}f & 0 \end{pmatrix}dz$$

$$+ *d\omega \begin{pmatrix} 0 & -1 & 0 \\ 1 & 0 & 0 \\ 0 & 0 & 0 \end{pmatrix}$$

$$+ \frac{\lambda}{4}\begin{pmatrix} 0 & 0 & -e^\omega - e^{-\omega}\bar{f} \\ 0 & 0 & -ie^\omega + ie^{-\omega}\bar{f} \\ e^\omega + e^{-\omega}\bar{f} & ie^\omega - ie^{-\omega}\bar{f} & 0 \end{pmatrix}d\bar{z}.$$

Setting $\lambda^{-1}\alpha_1'$, α_0, and $\lambda\alpha_1''$ the first, second and third of the terms on the right hand side, respectively, this gives the splitting

$$\alpha_\lambda = \lambda^{-1}\alpha_1' + \alpha_0 + \lambda\alpha_1'',$$

where $\alpha_1'' = \overline{\alpha_1'}$. We observe that the matrix α_0 has non-vanishing entries according to the pattern

$$\begin{pmatrix} 0 & * & 0 \\ * & 0 & 0 \\ 0 & 0 & 0 \end{pmatrix},$$

and α_1', α_1'' according to

$$\begin{pmatrix} 0 & 0 & * \\ 0 & 0 & * \\ * & * & 0 \end{pmatrix}.$$

In the particular case where $f = -1$, α_1' takes the form

$$\alpha_1' = \frac{1}{2} \begin{pmatrix} 0 & 0 & -\sinh\omega \\ 0 & 0 & i\cosh\omega \\ \sinh\omega & -i\cosh\omega & 0 \end{pmatrix} dz,$$

and α_1'' the conjugate of this.

Interpretation. Let $\alpha = \alpha_\lambda|_{\lambda=1}$. Then we have

   i) the splitting $\alpha = \alpha_0 + \alpha_1$, where α_0 has non-zero entries only in the first two rows of the first two columns and α_1 in the last row and in the last column,

   ii) the further splitting $\alpha_1 = \alpha_1' + \alpha_1''$, where $\alpha_1' = \alpha_1\left(\frac{\partial}{\partial z}\right) dz$ and $\alpha_1'' = \overline{\alpha_1'}$,

   iii) the deformation

$$\alpha_\lambda = \lambda^{-1}\alpha_1' + \alpha_0 + \lambda\alpha_1''.$$

Thus i)–iii) give an algorithm that computes α_λ from α without bothering with A_λ, or even F_λ. The following shows that we need just a harmonic map to start this construction.

Theorem 5.1 *Let $G = (g_1, g_2, u)\colon \Omega \to SO(3, \mathbb{R})$ and $\alpha := G^{-1}dG$. Suppose that α_λ is constructed from α according to i)–iii). If $u\colon \Omega \to S^2$ is a harmonic map, then*

$$d\alpha_\lambda + \alpha_\lambda \wedge \alpha_\lambda = 0,$$

and vice versa.

To prove the one implication, recall that any harmonic map into S^2 is the Gauss map of a CMC immersion. Thus α_λ belongs to the corresponding conjugate family of CMC immersions, and satisfies therefore the zero curvature condition. The other implication will be proved later.

Remark In the derivation of these equations using a complex parameters, we have followed a point of view close to [10].

Note that, beside CMC surfaces, there exist other examples of family of immersions which have the same first fundamental form and the same mean curvature function and which are geometrically different. They are solutions of the so-called *Bonnet problem*, see [13] for a review.

Recovering the CMC immersion from the S^1-family of harmonic maps (Bobenko-Sym's formula)

Once having constructed α_λ as above, we can get the CMC immersions belonging to α_λ back as follows.

Proposition 5.1 *Let $G_\lambda : \Omega \to \mathrm{SO}(3, \mathbb{R})$ be a solution of*

$$\begin{cases} dG_\lambda &= G_\lambda \alpha_\lambda, \\ G_\lambda(z_0) &= \mathbf{1}_3, \end{cases} \tag{5.8}$$

where $\alpha_\lambda = \lambda^{-1} \alpha_1' + \alpha_0 + \lambda \alpha_1''$, and

$$\alpha_1' = \alpha_1 \left(\frac{\partial}{\partial z} \right) dz, \quad \alpha_1'' = \overline{\alpha_1'}.$$

Moreover, let $\alpha_0, \alpha_1 \in \mathrm{so}(3, \mathbb{R})$ have non-vanishing entries in the two leftmost columns of the two uppermost rows, and in the last row and the last column, respectively. Let $u : \Omega \to S^2$ be the last column of G_1, and suppose u is harmonic. Write

$$\left. \frac{dG_{e^{it}}}{dt} \right|_{t=0} G_1^{-1} = \begin{pmatrix} 0 & B^3 & -B^2 \\ -B^3 & 0 & B^1 \\ B^2 & -B^1 & 0 \end{pmatrix}.$$

Then $B \pm u$ are weakly conformal CMC immersions and have u as their Gauss map.

Proof. Recall (see Chapter 2) that we can construct two weakly conformal CMC immersions $B \pm u$ belonging to u by setting

$$dB = u \times \frac{\partial u}{\partial y} \, dx - u \times \frac{\partial u}{\partial x} \, dy. \tag{5.9}$$

All we have to prove therefore is that these two constructions yield the same maps.

Write $\lambda = e^{it}$, for real t (with small $|t|$). Then $\lambda = 1 + it + o(t)$, and

$$\alpha_\lambda = \alpha + t(-i\alpha_1' + i\alpha_1'') + o(t),$$

where $\alpha = \alpha_\lambda|_{\lambda=1}$. But it is easily computed that $-i\alpha_1' + i\alpha_1'' = *\alpha_1$. So

$$\alpha_\lambda = \alpha + t(*\alpha_1) + o(t).$$

Now set $b = ((\frac{d}{dt} G_{e^{it}})|_{t=0}) G_1^{-1}$, so that

$$G_\lambda = G + tbG + o(t).$$

(We abbreviate $G = G_1$.) The condition (5.8) then implies

$$dG + t(dbG + bdG) = (G + tbG)(\alpha + t(*\alpha_1)) + o(t),$$

which is equivalent to the system

$$\begin{cases} dG &= G\alpha, \\ dbG + bdG &= bG\alpha + G(*\alpha_1). \end{cases} \tag{5.10}$$

From system (5.10), using the first equation to eliminate dG in the second one, we conclude that

$$db = *(G\alpha_1 G^{-1}). \tag{5.11}$$

Now, write

$$\alpha_1 = \begin{pmatrix} 0 & 0 & \alpha_3^1 \\ 0 & 0 & \alpha_3^2 \\ \alpha_1^3 & \alpha_2^3 & 0 \end{pmatrix}, \quad G = (e_1, e_2, u).$$

Then (5.11) can be written in the form

$$db = *((\alpha_3^1 e_1 + \alpha_3^2 e_2)^t u - u^t(\alpha_3^1 e_1 + \alpha_3^2 e_2)),$$

whereas

$$du = \alpha_3^1 e_1 + \alpha_3^2 e_2,$$

according to (5.8). This means that

$$db = *((du)^t u - u^t(du)).$$

Comparing this with (5.9), we see that b corresponds—up to a constant—to B, and the proof is thus complete. $\qquad\qquad\qquad\qquad\qquad\qquad\qquad\qquad\Box$

5.1 Appendix

We give here a proof of the following result.

Lemma 5.2 *Let Ω be a simply connected open subset of $\mathbb{R}^m$, and let*

$$A \in \mathcal{C}^1(\Omega, T^*\mathbb{R}^m \otimes \mathrm{gl}(n, \mathbb{R}))$$

be a 1-form on Ω with coefficients in the set of matrices $\mathrm{gl}(n, \mathbb{R}) \simeq M(n, \mathbb{R})$ (the Lie algebra of $\mathrm{GL}(n, \mathbb{R})$) such that

$$dA + A \wedge A = 0,$$

i.e., writing $A = \sum_{\alpha=1}^{m} A_\alpha dx^\alpha$,

$$\sum_{1 \leq \alpha < \beta \leq m} \left(\frac{\partial A_\beta}{\partial x^\alpha} - \frac{\partial A_\alpha}{\partial x^\beta} + [A_\alpha, A_\beta] \right) dx^\alpha \wedge dx^\beta = 0.$$

Then, given any $F_0 \in \mathrm{GL}(n, \mathbb{R})$, $x_0 \in \Omega$, $\exists! F \in \mathcal{C}^2(\Omega, \mathrm{GL}(n, \mathbb{R}))$ such that

$$F(x_0) = F_0 \text{ and } dF = FA.$$

Proof.
Step 1: Local existence. We define the m following vector fields on $\Omega \times \mathrm{GL}(n, \mathbb{R})$,

$$X_\alpha(x, M) := \frac{\partial}{\partial x^\alpha} + (MA_\alpha(x)) \bullet \frac{\partial}{\partial M},$$

where we denote

$$(MA_\alpha(x)) \bullet \frac{\partial}{\partial M} := \sum_{1 \leq i,j,k \leq n} M_j^i A_\alpha(x)_k^j \frac{\partial}{\partial M_k^i}.$$

Notice that

$$\left[(MA_\alpha(x)) \bullet \frac{\partial}{\partial M} \right] \left[(MA_\beta(x)) \bullet \frac{\partial}{\partial M} \right] = (MA_\alpha(x)A_\beta(x)) \bullet \frac{\partial}{\partial M}.$$

Then

$$[X_\alpha, X_\beta](x, M) = \left(M \left(\frac{\partial A_\beta}{\partial x^\alpha} - \frac{\partial A_\alpha}{\partial x^\beta} + [A_\alpha, A_\beta] \right) \right) \bullet \frac{\partial}{\partial M} = 0.$$

Thus these vector fields commute and there exists an open subset U of $\mathbb{R}^m$ and a map

$$\begin{array}{rccc} \Phi: & U & \longrightarrow & \Omega \times \mathrm{GL}(n, \mathbb{R}) \\ & t & \longrightarrow & (\psi(t), F(t)), \end{array}$$

such that for all $1 \leq \alpha \leq m$, $\frac{\partial \Psi}{\partial t^\alpha}(t) = X_\alpha(\Psi(t))$. By a suitable redefinition of t, one sees easily that $\psi(t) = t = x$ and that $\frac{\partial F}{\partial x^\alpha} = FA_\alpha$.

Step 2: Local "uniqueness", up to a constant. Let F and $\tilde{F}$ be two local solutions. On their overlapping domains, if $dF = FA$ and $d\tilde{F} = \tilde{F}A$, then one compute easily that $d(\tilde{F}F^{-1}) = 0$ and thus $\tilde{F} = CF$, for some constant $C \in \mathrm{GL}(n, \mathbb{R})$.

Step 3: Global result. Follows from the fact that Ω is simply connected. □

We remark that the same result holds also with complex matrices (i.e. by replacing $\mathrm{GL}(n, \mathbb{R})$ by $\mathrm{GL}(n, \mathbb{C})$). Also, if $\mathfrak{G}$ is any Lie sub-group of $\mathrm{GL}(n, \mathbb{R})$, with Lie algebra $\mathfrak{g}$, then it is easy to check that if $F_0 \in \mathfrak{G}$ and A has its coefficients in $\mathfrak{g}$, then F takes its values in $\mathfrak{G}$. Examples of such situations are $\mathrm{SO}(n)$ with Lie algebra $\mathrm{so}(n)$, $\mathrm{SL}(n)$ with Lie algebra $\mathrm{sl}(n)$, $\mathrm{SU}(n)$ with Lie algebra $\mathrm{su}(n)$, $\mathrm{U}(n)$ with Lie algebra $\mathrm{u}(n)$.

6 Elementary twistor theory for harmonic maps

Twistor theory as we are going to see it was first developed by E. Calabi [25] in 1967. Other contributions were made—among others—by J. Eells and J. Wood in 1982 [34], F. Burstall and J. H. Rawnsley in 1986 [23], and K. Uhlenbeck in 1989 [82]. We are going to consider harmonic maps $u: S^2 \to S^n \subset \mathbb{R}^{n+1}$, i. e. maps satisfying

$$\Delta u + u|du|^2 = 0,$$

which is equivalent to $\Delta u \parallel u$. In contrast to Chapter 4, where we considered the Hopf differential

$$\left(\frac{\partial u}{\partial z}, \frac{\partial u}{\partial z} \right) (dz)^2,$$

we will also use derivatives of u of higher order. Let us first introduce some notations. We write

$$\partial = \frac{\partial}{\partial z}, \quad \bar{\partial} = \frac{\partial}{\partial \bar{z}}.$$

If $V, W \in \mathbb{C}^{n+1}$, then the complexification of the scalar product on $\mathbb{R}^{n+1}$ is

$$(V, W) = \sum_{i=1}^{n+1} V^i W^i,$$

and the standard Hermitian scalar product is denoted

$$\langle V, W \rangle = \sum_{i=1}^{n+1} V^i \overline{W}^i.$$

Consider the expressions

$$\eta_{\alpha\beta} = (\partial^\alpha u, \partial^\beta u) = \langle \partial^\alpha u, \bar{\partial}^\beta u \rangle, \quad \alpha, \beta \geq 0,$$

in some local coordinate $z \in \Omega$ on S^2. Note that these are no tensors. If we have a holomorphic change of variables $\phi: \Omega_2 \to \Omega_1$, where Ω_1, Ω_2 are open subsets of $\mathbb{C}$, then η_{21} (for instance) transforms as follows by ϕ:

$$(\partial^2(u \circ \phi), \partial(u \circ \phi)) = ((\partial^2 u, \partial u) \circ \phi)(\partial\phi)^3 + ((\partial u, \partial u) \circ \phi)(\partial^2\phi)(\partial\phi).$$

So we have in general

$$\phi^*(\eta_{\alpha\beta}) \neq (\eta_{\alpha\beta} \circ \phi)(\partial\phi)^{\alpha+\beta}.$$

However, equality holds up to expressions containing $\eta_{\alpha'\beta'}$ with $\alpha' + \beta' < \alpha + \beta$. We have seen this for η_{21}; in general, it follows by induction. Furthermore, the following holds.

Proposition 6.1 *Let* $u\colon S^2 \to S^n$ *be harmonic. Then, in all holomorphic local coordinates* z, *and for all* $\alpha, \beta \geq 0$, *such that* $\alpha + \beta \geq 1$, *we have* $\eta_{\alpha\beta} = 0$.

Definition 6.1 *We say that* $u\colon \Omega \to S^n$ *is isotropic, if and only if* $\eta_{\alpha\beta} = 0$ *for all* $\alpha, \beta \geq 0$ *satisfying* $\alpha + \beta \geq 1$.

Proof of Proposition 6.1. We will apply induction on $\alpha + \beta$. The claim is clear for $\alpha + \beta = 1$, since

$$\eta_{01} = \eta_{10} = (\partial u, u) = 0.$$

Assume now that $\gamma \in \mathbb{N}$ is given, and that $\eta_{\alpha\beta} = 0$ for all α, β such that $1 \leq \alpha + \beta \leq \gamma$.

Step 1: $\eta_{\gamma 1}(dz)^{\gamma+1}$ *is a tensor.* Let ϕ be a holomorphic change of variables. Compute

$(\partial^\gamma(u \circ \phi), \partial(u \circ \phi))$

$$= \left((\partial^\gamma u \circ \phi)(\partial\phi)^\gamma + \sum_{k=1}^{\gamma-1} P_k^\gamma(\partial\phi, \partial^2\phi, \ldots, \partial^{\gamma+1-k}\phi)(\partial^k u \circ \phi), (\partial u \circ \phi)(\partial\phi) \right)$$

$$= ((\partial^\gamma u, \partial u) \circ \phi)(\partial\phi)^{\gamma+1} + \sum_{k=1}^{\gamma-1} P_k^\gamma(\partial\phi, \partial^2\phi, \ldots, \partial^{\gamma+1-k}\phi)(\eta_{k1} \circ \phi)(\partial\phi)$$

$$= (\eta_{\gamma 1} \circ \phi)(\partial\phi)^{\gamma+1}$$

for some polynomials P_k^γ. This proves the claim.

Step 2: $\eta_{\gamma 1}(dz)^{\gamma+1}$ *is holomorphic.* We use the harmonic map equation

$$\partial\bar\partial u = \frac{1}{4}\Delta u = -\frac{1}{4}u|du|^2.$$

It implies

$$\bar\partial(\partial^\gamma u, \partial u) = (\partial^{\gamma-1}(\partial\bar\partial u), \partial u) + (\partial^\gamma u, \bar\partial\partial u)$$

$$= -\frac{1}{4}[(\partial^{\gamma-1}(u|du|^2), \partial u) + (\partial^\gamma u, u|du|^2)].$$

The expression $\partial^{\gamma-1}(u|du|^2)$ gives a linear combination of $u, \partial u, \ldots, \partial^{\gamma-1}u$ with coefficients that consist of the derivatives of $|du|^2$. Hence $\bar\partial(\partial^\gamma u, \partial u) = 0$, and Step 2 is complete.

Step 3: Use Liouville's theorem to conclude $\eta_{\gamma 1} = 0$.

Step 4: Completion of the proof. Note that for all $\alpha + \beta = \gamma$, we have $\eta_{\alpha\beta} = 0$ and hence

$$0 = \partial\eta_{\alpha\beta} = \partial(\partial^\alpha u, \partial^\beta u) = (\partial^{\alpha+1}u, \partial^\beta u) + (\partial^\alpha u, \partial^{\beta+1}u) = \eta_{\alpha+1,\beta} + \eta_{\alpha,\beta+1}.$$

So we can conclude from $\eta_{\gamma 1} = 0$, that

$$\eta_{\gamma-1,2} = 0, \eta_{\gamma-2,3} = 0, \ldots, \eta_{0,\gamma+1} = 0,$$

which proves the proposition. $\qquad\qquad\qquad\qquad\qquad\qquad\qquad\qquad\qquad\qquad$ $\square$

Isotropy is a strong property, but not easy to work with. However, with a little more preparations, we will finally be able to derive the following consequences. If $u \colon S^2 \to S^n$ is so that $u(S^2)$ is not contained in a hyperplane in $\mathbb{R}^{n+1}$ (such maps are called "full"), then

  i) n is even (see lemma 6.2),

  ii) there is some holomorphic data associated to u (Calabi [25], Eells–Wood [34]),

 iii) the energy of u is quantized (Calabi [25]).

We assume the following hypotheses: $u \colon \Omega \to S^n$ is a full, isotropic map. We define, for any point $z \in \Omega$,

$$\begin{cases} \theta'_\alpha &= \text{Span}(\partial u, \ldots, \partial^\alpha u), \quad \alpha \in \mathbb{N}, \\ \theta''_\beta &= \text{Span}(\bar\partial u, \ldots, \bar\partial^\beta u), \quad \beta \in \mathbb{N}, \end{cases}$$

in the sense of complex vector spaces (i. e. $\theta'_\alpha, \theta''_\beta$ are in the Grassmannian $\text{Gr}(\mathbb{C}^{n+1})$). Note that, for all α, β, the conditions $\eta_{\mu\nu} = 0$, $\forall \mu, \nu$ such that $\mu \le \alpha$ and $\nu \le \beta$ mean that

$$\theta'_\alpha \perp \theta''_\beta.$$

Here, $\perp$ means perpendicular with respect to $\langle \cdot, \cdot \rangle$. Finally, we define

$$\Theta_q = \text{Span}\{\partial^\alpha \bar\partial^\beta u \colon 0 \le \alpha + \beta \le q\}, \quad q \ge 0.$$

Technical lemmas

We will state here a few properties of the spaces just defined. Some of the corresponding proofs will be given immediately, some in the appendix of this chapter.

Lemma 6.1 *Assume that $u \colon \Omega \to S^n$ is harmonic. Then*

$$\Theta_q = \theta'_q + \theta''_q + \mathbb{C}u$$

for all q.

Lemma 6.2 *Assume that $u \colon \Omega \to S^n$ is full, isotropic, and harmonic, and that Ω is connected. Let*

$$r := \max_{\substack{z \in \Omega \\ \alpha \ge 0}} \dim \theta'_\alpha.$$

Then $2r = n$.

A consequence of this result is that n is necessarily even. We shall use the notation $n := 2r$ in the rest of this Chapter.

Proof of $2r \leq n$. By the definition of r, there exists a point $z \in \Omega$, and there are $\gamma_1, \ldots, \gamma_r \geq 1$, such that $(\partial^{\gamma_1} u, \ldots, \partial^{\gamma_r} u)(z)$ is of rank r. By isotropy, we have

$$\langle \partial^{\gamma_i} u, \bar{\partial}^{\gamma_j} u \rangle = 0, \quad 1 \leq i, j \leq r,$$

and

$$\langle u, \partial^{\gamma_i} u \rangle = \langle u, \bar{\partial}^{\gamma_i} u \rangle = 0, \quad 1 \leq i \leq r.$$

The spaces

$$\mathrm{Span}(\partial^{\gamma_1} u, \ldots, \partial^{\gamma_r} u)(z), \quad \mathrm{Span}(\bar{\partial}^{\gamma_1} u, \ldots, \bar{\partial}^{\gamma_r} u)(z), \quad \text{and} \quad \mathbb{C}u$$

are thus mutually perpendicular. The direct sum

$$\mathrm{Span}(\partial^{\gamma_1} u, \ldots, \partial^{\gamma_r} u)(z) \oplus \mathrm{Span}(\bar{\partial}^{\gamma_1} u, \ldots, \bar{\partial}^{\gamma_r} u)(z) \oplus \mathbb{C}u$$

is a $(2r + 1)$-dimensional subspace of $\mathbb{C}^{n+1}$. Hence $2r \leq n$. $\square$

Lemma 6.3 *Let* $u: \Omega \to S^n$ *be real analytic. Set*

$$B := \{z \in \Omega : \dim \theta'_r < r\}.$$

Then $\Omega \backslash B$ *is non-empty.*

Lemma 6.4 *Let* $u: \Omega \to S^n$ *be a full, isotropic, harmonic map on a connected domain* Ω. *Let* B *be defined as in the previous lemma. Define*

$$\phi_\alpha(z) = \theta'_\alpha(z) \in \mathrm{Gr}_\alpha(\mathbb{C}^{n+1}), \quad 1 \leq \alpha \leq r,$$

on $\Omega \backslash B$. *Then* ϕ_α *has a unique real analytic extension on* Ω, *and* B *is composed of isolated points.*

Proof. Let

$$w_\alpha(z) = \partial u \wedge \partial^2 u \wedge \ldots \wedge \partial^\alpha u: \Omega \to \Lambda^\alpha \mathbb{C}^{n+1}.$$

Then B is the set of all zeros of w_r. We want to show that w_α behaves like a holomorphic function in the following sense: if $w_\alpha(z_0) = 0$, then

$$w_\alpha(z) = (z - z_0)^k W_\alpha(z)$$

in a neighbourhood of z_0, where $k \in \mathbb{N}$ and W_α is real analytic with $W_\alpha(z_0) \neq 0$. Then w_α has of course isolated zeros, and W_α provides locally an extension of ϕ_α.

For any $z \in \Omega$, we have $w_\alpha(z) \in \Lambda^\alpha(u^\perp)$, since $\partial u, \ldots, \partial^\alpha u \perp u$. Consider the orthogonal projection $P(z) \colon \mathbb{C}^{n+1} \to u^\perp(z)$ and the unique linear mapping $P_\alpha(z) \colon \Lambda^\alpha \mathbb{C}^{n+1} \to \Lambda^\alpha \mathbb{C}^{n+1}$ such that for all $v_1, \ldots, v_\alpha \in \mathbb{C}^{n+1}$,

$$P_\alpha(z)(v_1 \wedge \ldots \wedge v_\alpha) = P(z)v_1 \wedge \ldots \wedge P(z)v_\alpha.$$

Of course, we have

$$P_\alpha(z)w_\alpha(z) = w_\alpha(z). \tag{6.1}$$

Now, compute

$$
\begin{aligned}
\bar{\partial} w_\alpha &= \bar{\partial}\partial u \wedge \partial^2 u \wedge \ldots \wedge \partial^\alpha u + \partial u \wedge \bar{\partial}\partial^2 u \wedge \partial^3 u \wedge \ldots \wedge \partial^\alpha u \\
&\quad + \ldots + \partial u \wedge \ldots \wedge \partial^{\alpha-1} u \wedge \bar{\partial}\partial^\alpha u \\
&= -\frac{1}{4}[(u \wedge \partial^2 u \wedge \ldots \wedge \partial^\alpha u)|du|^2 + (\partial u \wedge u \wedge \partial^3 u \wedge \ldots \wedge \partial^\alpha u)\partial|du|^2 \\
&\quad + \ldots + (\partial u \wedge \ldots \wedge \partial^{\alpha-1} u \wedge u)\partial^{\alpha-1}|du|^2].
\end{aligned}
$$

Hence $P_\alpha(z)\bar{\partial}w_\alpha(z) = 0$. Derivating (6.1) now yields

$$\bar{\partial}w_\alpha = \bar{\partial}P_\alpha w_\alpha + P_\alpha \bar{\partial} w_\alpha = \bar{\partial}P_\alpha.w_\alpha.$$

Lemma 6.5 in the appendix will prove that this implies indeed the claim. $\square$

Conclusion

Suppose $u \colon \Omega \to S^n$ is a full, isotropic, harmonic map on a connected domain. The lemmas of the previous section allow us to construct some holomorphic data that are associated to u. To this end, let

$$l(z) = (\theta'_r \oplus \mathbb{C}u \oplus \theta''_{r-1})^\perp.$$

This is a 1-dimensional subspace of $\mathbb{C}^{n+1}$, we may therefore regard it as an element of $\mathbb{C}P^n$.

Claim The map $l \colon \Omega \to \mathbb{C}P^n$ is holomorphic.

Proof. Consider

$$\gamma = u \wedge \partial u \wedge \ldots \wedge \partial^r u \wedge \bar{\partial}u \wedge \ldots \wedge \bar{\partial}^{r-1}u.$$

Use the harmonic map equation to compute

$$\partial\gamma = u \wedge \partial u \wedge \ldots \wedge \partial^{r-1} u \wedge \partial^{r+1} u \wedge \bar{\partial}u \wedge \ldots \wedge \bar{\partial}^{r-1}.$$

By the choice of r, we may write $\partial^{r+1} u = V + \mu \, \partial^r u$, where $\mu \in \mathbb{C}$ and $V \in \mathrm{Span}(\partial u, \ldots \partial^{r-1} u)$. So we get

$$\partial \gamma = \mu \gamma.$$

Moreover, μ depends analytically on $z \in \Omega$. There exists a vector ψ in the line l (depending on z), such that for all $X \in \mathbb{C}^{n+1}$, we have

$$\gamma \wedge X = \langle X, \psi \rangle \, \epsilon_1 \wedge \ldots \wedge \epsilon_{n+1},$$

where $\epsilon_1, \ldots \epsilon_{n+1}$ constitute the standard basis of $\mathbb{C}^{n+1}$. Then we get

$$\mu \langle X, \psi \rangle \, \epsilon_1 \wedge \ldots \wedge \epsilon_{n+1} = \mu \gamma \wedge X = \partial(\gamma \wedge X) = \langle X, \bar{\partial} \psi \rangle \, \epsilon_1 \wedge \ldots \wedge \epsilon_{n+1}.$$

Hence $\bar{\partial} \psi = \bar{\mu} \psi$. Let $z_0 \in \Omega$, and choose a simply connected open neighbourhood Ω' of z_0. Let $\lambda \colon \Omega' \to \mathbb{C}^*$ be a solution of

$$\bar{\partial} \lambda + \bar{\mu} \lambda = 0, \tag{6.2}$$

and set $\phi = \lambda \psi$. Obviously ϕ is holomorphic. But ϕ is a representation of l, so also l is holomorphic at z_0. Since z_0 can be chosen arbitrarily, l is holomorphic in Ω. $\qquad\square$

Remark To solve (6.2), first find a solution of

$$\bar{\partial} \partial g = \frac{1}{4} \Delta g = \bar{\partial} \mu.$$

Then set

$$f(z) = \int_{z_0}^z (\partial g - \mu) \, d\zeta - g(z).$$

Note that this integral makes sense, since $\bar{\partial}(\partial g - \mu) = 0$. The function f satisfies

$$\partial f = -\mu.$$

Hence if we set $\lambda = \overline{e^f}$, then

$$\bar{\partial} \lambda + \bar{\mu} \lambda = \overline{(\partial f + \mu) e^f} = 0.$$

Example (for $r = 1$.) If $u : \Omega \longrightarrow S^2$ is a full isotropic harmonic map, then $\gamma = u \wedge \partial u$ and $l = (\mathbb{C}u + \mathbb{C}\partial u)^\perp = \mathbb{C}\bar{\partial}u$. We know that we can write

$$u = \pm \frac{1}{1 + |v|^2} \begin{pmatrix} v + \bar{v} \\ -i(v - \bar{v}) \\ 1 - |v|^2 \end{pmatrix},$$

where $\pm$ is $+$ if u is holomorphic, and $-$ if u is antiholomorphic. Then

$$\overline{\partial} u = \pm \frac{i}{(1+|v|^2)^2} \frac{\partial \overline{v}}{\partial z} \begin{pmatrix} i(v^2 - 1) \\ v^2 + 1 \\ 2iv \end{pmatrix},$$

so that, in homogeneous coordinates, $l(z) = [i(v^2 - 1) : v^2 + 1 : 2iv]$. This is a holomorphic curve into the quadric

$$\{[z^1 : z^2 : z^3] \in \mathbb{C}\mathrm{P}^2 : (z^1)^2 + (z^2)^2 + (z^3)^2 = 0\} \simeq \mathbb{C}\mathrm{P}^1.$$

Another point of view

Here we still denote $n = 2r$. Consider the complex submanifold

$$Z_0 = \{V \in \mathrm{Gr}_r(\mathbb{C}^{n+1}) : V \perp \overline{V}\}$$

of $\mathrm{Gr}_r(\mathbb{C}^{n+1})$. We call a subspace $V \subset \mathbb{C}^{n+1}$ that satisfies $V \perp \overline{V}$ isotropic. The space Z_0 has the following properties.

- Z_0 is fibered over S^n, i. e. there exists a projection $\Pi \colon Z_0 \to S^n$. It can be constructed as follows.

  Let $V \in Z_0$. By isotropy, $V \oplus \overline{V}$ is a subspace of complex dimension n of $\mathbb{C}^{n+1}$. Moreover, it is invariant by conjugation. The complex line $(V \oplus \overline{V})^\perp$ is also invariant by conjugation. Thus we may choose a real vector $y \in \mathbb{R}^{n+1}$ such that $|y| = 1$ and $(V \oplus \overline{V})^\perp = \mathbb{C}y$. The vector y is obtained from any $x \in (V \oplus \overline{V})^\perp$ by setting either $y = \frac{x+\bar{x}}{|x+\bar{x}|}$ or $y = \frac{i(x-\bar{x})}{|x-\bar{x}|}$. Taking into account that the decomposition $\mathbb{C}^{n+1} = V \oplus \overline{V} \oplus \mathbb{C}y$ gives an orientation of $\mathbb{C}^{n+1}$ (or $\mathbb{R}^{n+1}$), we can make y uniquely determined by prescribing the orientation. More precisely, take an orthonormal basis $(v_1, \ldots, v_r)$ of V. Set $e_{2k-1} = \frac{1}{\sqrt{2}}(v_k + \bar{v}_k)$, $e_{2k} = \frac{1}{\sqrt{2}}i(v_k - \bar{v}_k)$, for $k = 1, \ldots, r$. Choose y so that $(e_1, \ldots, e_{2n}, y)$ is an orthonormal, positively oriented basis of $\mathbb{R}^{n+1}$. Now set

$$\Pi(V) = y,$$

  and this gives the fibration of Z_0.
- $Z_0 \simeq \mathrm{SO}(n+1)/\mathrm{U}(r)$, where

$$\mathrm{U}(r) \simeq \{\text{Hermitian bases of } V\}$$

for a given r-dimensional space V, and

$$\mathrm{SO}(n+1) \simeq \{\text{oriented orthonormal bases of } \mathbb{R}^{n+1}\}.$$

Note that $\mathrm{U}(r)$ can be seen as a subgroup of $\mathrm{SO}(n)$ and hence of $\mathrm{SO}(n+1)$ by the identification of $\mathbb{R}^n$ with $\mathbb{C}^r$. The isomorphism $Z_0 \simeq \mathrm{SO}(n+1)/\mathrm{U}(r)$ is given by the construction of an oriented basis of $\mathbb{R}^{n+1}$ as described above. Of course, this basis is only uniquely determined up to the choice of the basis $(v_1, \dots, v_r)$ of V.

- The space $Z = Z_0 \sqcup Z_0$ is the sub-bundle of isotropic r-spaces in $TS^{n+1} \otimes \mathbb{C}$, for we can identify $(T_x S^{n+1} \otimes \mathbb{C}) \sqcup (T_{-x} S^{n+1} \otimes \mathbb{C})$ with two copies of an element of the Grassmann bundle, for any $x \in S^{n+1}$. We call Z the *twistor bundle*.

Now let $u \colon \Omega \to S^n$ be a full, isotropic, harmonic map. Consider the map

$$V \colon \Omega \to \mathrm{Gr}_r(\mathbb{C}^{n+1}),$$

defined by

$$V = \mathrm{Span}(\bar{\partial} u, \dots, \bar{\partial}^r u).$$

Then we have $\overline{V} = \mathrm{Span}(\partial u, \dots, \partial^r u)$. This is perpendicular to V, hence V is a map into Z_0. Moreover, we have $\bar{\partial} V \subset V$. With the aid of the harmonic map equation, it is easy to see that $\partial V \subset V \oplus \mathbb{C}u$.

Definition 6.2 *A map $V \colon \Omega \to Z_0$ is called holomorphic, if and only if $\bar{\partial} V \subset V$. It is horizontal, if and only if $dV \subset V \oplus \mathbb{C}y$, where $y = \Pi(V)$.*

So $V \colon \Omega \to Z_0$ is both holomorphic and horizontal, if it is obtained from a full, isotropic, harmonic map as above. But the converse is also true.

Proposition 6.2 *Let $V \colon \Omega \to Z_0$ be holomorphic and horizontal. Then $u = \Pi(V)$ is harmonic and isotropic.*

Proof. For all $v \in V$, we have $\langle v, u \rangle = 0$. Hence $\langle \bar{\partial} v, u \rangle + \langle v, \partial u \rangle = 0$. But we have also

$$\bar{\partial} v \in \bar{\partial} V \subset V \perp \mathbb{C}u.$$

So $\langle v, \partial u \rangle = 0$, which means $\partial u \perp V$. Since also $\partial u \perp u$ (recall $|u| = 1$), we have in fact $\partial u \in \overline{V}$. Derivating with respect to $\bar{z}$ now gives

$$\bar{\partial} \partial u \in \bar{\partial} \overline{V} \subset \overline{V} + \mathbb{C}u.$$

The same can be done with ∂ and $\bar{\partial}$ exchanged. Then we get

$$\partial \bar{\partial} u \in V + \mathbb{C}u.$$

The orthogonality of V and $\overline{V}$ implies therefore that $\Delta u = 4 \partial \bar{\partial} u \parallel u$, and u is harmonic. From the fact $\bar{\partial} u \in V$ we conclude that $\bar{\partial}^\alpha u \in V$ for all α, for V is holomorphic. Hence $\partial^\alpha u \in \overline{V}$, and the isotropy of u follows. $\qquad \square$

Remark A slight generalization of this theory allows to associate holomorphic datas to any harmonic maps $S^2 \longrightarrow \mathbb{C}P^n$ and conversely to construct all such harmonic maps starting from holomorphic curves in to $\mathbb{C}P^n$ (the suitable twistor space here) [34]. More precisely we can associate to each full holomorphic $\phi : S^2 \longrightarrow \mathbb{C}P^n$ a sequence of $n + 1$ harmonic maps into $\mathbb{C}P^n$ named a *harmonic sequence* $\phi_0, \phi_1, \ldots, \phi_n$, where in particular $\phi_0 = \phi$ is holomorphic, ϕ_n is antiholomorphic and the other ϕ_j's "interpolate" between ϕ_0 and ϕ_n. Focusing on the relation between each element of a harmonic sequence leads to notions which may be further generalized to non isotropic harmonic maps and is connected with the Toda equation, as investigated by J. Bolton and L. Woodward in [16].

6.1   Appendix

We complete here the proofs concerning the analysis of full isotropic harmonic maps from an open subset domain $\Omega \subset \mathbb{C}$ into S^n. First we prove

Lemma 6.1 *Let $u : \Omega \longrightarrow S^n$ be a harmonic map and $\theta'_\alpha = \mathbb{C}\partial u + \cdots + \mathbb{C}\partial^\alpha u$, $\theta''_\beta = \mathbb{C}\overline{\partial} u + \cdots + \mathbb{C}\overline{\partial}^\beta u$, and $\Theta_q = \mathrm{Span}\{\partial^\alpha \overline{\partial}^\beta u : 0 \le \alpha + \beta \le q\}$. Then*

$$\Theta_q = \theta'_q + \theta''_q + \mathbb{C}u.$$

Proof. The non-obvious part is to prove the inclusion $\subset$. We prove it by induction on q. The result is clear for $q = 0$. Assume that that the claim is true for some $q \in \mathbb{N}$. Any vector V in Θ_{q+1} is a sum $V = V_q + \sum_{\alpha=0}^{q+1} c_\alpha \partial^\alpha \overline{\partial}^{q+1-\alpha} u$, where $V_q \in \Theta_q = \theta'_q + \theta''_q + \mathbb{C}u$. So we need only to prove the claim for each $\partial^\alpha \overline{\partial}^{q+1-\alpha} u$. Exclude the easy case $\alpha = 0$ or $q + 1$, then

$$\partial^\alpha \overline{\partial}^{q+1-\alpha} u = \frac{1}{4} \partial^{\alpha-1} \overline{\partial}^{q-\alpha} (\Delta u) = -\frac{1}{4} \partial^{\alpha-1} \overline{\partial}^{q-\alpha} (u|du|^2) \in \Theta_{q-1},$$

by developing the last expression. This proves the Lemma. $\qquad\qquad\square$

Lemma 6.2 *Assume that Ω is connected. Let $u : \Omega \longrightarrow S^n$ be a full isotropic harmonic map and set*

$$r := \max_{z \in \Omega, \alpha \ge 0} \dim \theta'_\alpha.$$

Then

$$n = 2r.$$

Proof. Since we proved $n \ge 2r$ before, we show here that $n \le 2r$. Let us assume the contrary: $n > 2r$. Then, by Lemma 6.1, choosing any point z, $\dim \mathrm{Span}\{\partial^\alpha \overline{\partial}^\beta u(z) : 0 \le \alpha, \beta\} \le 2r + 1 < n + 1$. So there exists some complex hyperplane $H_z \subset \mathbb{C}^{n+1}$, such that $\mathrm{Span}\{\partial^\alpha \overline{\partial}^\beta u(z)/0 \leq \alpha, \beta\} \subset H_z$. Let $f_z : \mathbb{C}^{n+1} \longrightarrow \mathbb{C}$ a linear form of kernel H_z, we deduce that $f_z \circ u$ vanishes at infinite order at z and thus is identically zero by analycity, a contradiction. $\square$

Lemma 6.3 *Let $u : \Omega \longrightarrow S^n$ be a real analytic map and let*

$$B = \{z \in \Omega : \dim \theta'_r(z) < r\}.$$

Then $\Omega \backslash B \neq \emptyset$. In other words, there exists a point $z \in \Omega$ such that $\dim \theta'_r(z) = r$.

Proof. For $1 \leq \alpha \leq r$, let us denote $w_\alpha(z) = \partial u(z) \wedge \cdots \wedge \partial^\alpha u(z) \in \Lambda^\alpha \mathbb{C}^{n+1}$ and set $B_\alpha = \{z \in \Omega : w_\alpha(z) = 0\}$ and $B_0 = \emptyset$. Since all w_α's are continuous, all B_α's are closed subsets of Ω. Let

$$\alpha_0 := \sup\{\alpha : 0 \leq \alpha \leq r \text{ and } B_\alpha \neq \Omega\}.$$

Now let us assume that the claim in the Lemma is false. It means that $B_r = \Omega$ and thus that $\alpha_0 < r$. But on the nonempty open subset $\Omega \setminus B_{\alpha_0}$, $w_{\alpha_0+1} = 0$ and $w_{\alpha_0} \neq 0$, which implies that we may write $\partial^{\alpha_0+1} u = \sum_{\beta=1}^{\alpha_0} c_\beta \partial^\beta u$, for some smooth functions c_β (these coefficients are smooth because they are unique on $\Omega \backslash B_{\alpha_0}$). Derivating this relation with respect to z an arbitrary number of times, we may easily prove by induction that, on $\Omega \setminus B_{\alpha_0}$, $\partial^k u \in \theta'_{\alpha_0}$ for all $k \in \mathbb{N}^*$. To conclude, we consider maps $f_k = w_{\alpha_0} \wedge \partial^k u$. They vanish on $\Omega \setminus B_{\alpha_0}$ and are real analytic on Ω, thus vanish on Ω. It follows that $r \leq \alpha_0$, a contradiction. $\square$

We now prove the following analytic result, crucial for the proof of Lemma 6.4. It basically says that a holomorphic section of a complex vector bundle with connection behaves locally like a holomorphic function, in the sense that either it is identically zero, or it has isolated zeroes with finite multiplicity.

Lemma 6.5 ("local behaviour") *Let Ω be a connected open subset of $\mathbb{C}$ and $A : \Omega \longrightarrow GL(N, \mathbb{C})$ a smooth (C^∞) map into the set of linear (matrices) operators acting on $\mathbb{C}^N$. Assume that $w : \Omega \longrightarrow \mathbb{C}^N$ is a real analytic map, solution of*

$$\overline{\partial} w(z) = A(z) w(z) \text{ on } \Omega. \tag{6.3}$$

Set $B = \{z \in \Omega : w(z) = 0\}$. Then

- *either $B = \Omega$*
- *or B has isolated points and for any $z_0 \in B$, there exists some $p \in \mathbb{N}^*$ such that*

$$w(z) = (z - z_0)^p W(z) \text{ around } z_0,$$

where W is a real analytic and $W(z_0) \neq 0$.

Proof. Let $z_0 \in B$.

Step 1: We prove the following claim: Let $n \in \mathbb{N}$ such that for any $0 \leq k, l$ such that $k + l \leq n$,

$$\partial^k \overline{\partial}^l w(z_0) = 0.$$

Then, for any $0 \leq k, l$ such that $k + l \leq n$,

$$\partial^k \overline{\partial}^{l+1} w(z_0) = 0.$$

This is a consequence of (6.3), for

$$\partial^k \overline{\partial}^{l+1} w(z_0) = \partial^k \overline{\partial}^l (Aw)(z_0) = 0,$$

by developing the last expression.

Step 2: Now we suppose that, for all $n \in \mathbb{N}$, $\partial^n w(z_0) = 0$. Then using the claim of Step 1, one easily deduce by induction that, $\forall k, l \in \mathbb{N}$, $\partial^k \overline{\partial}^l w(z_0) = 0$. And thus, since w is real analytic and Ω is connected, one concludes that w vanishes everywhere.

Step 3: We assume that we are not in the case studied in Step 2. Hence there exists some $p \in \mathbb{N}^*$ such that $\forall 0 \leq k \leq p-1$, $\partial^k w(z_0) = 0$, but that $\partial^p w(z_0) \neq 0$. Then, arguing by induction from 0 to $n = p - 1$ as in Step 2, by using the claim of Step 1, we show that, for any $0 \leq k, l$ such that $k + l \leq p - 1$,

$$\partial^k \overline{\partial}^{l+1} w(z_0) = 0.$$

Hence a Taylor expansion at order p at z_0 gives:

$$w(z) = \frac{(z - z_0)^p}{p!} \partial^p w(z_0) + O(|z - z_0|^{p+1}),$$

and the conclusion follows. $\square$

7 Harmonic maps as an integrable system

7.1 Maps into spheres

The sphere as a quotient of Lie groups

We consider once more maps into S^n. Note that S^n can be identified with the group $\mathrm{SO}(n+1)/\mathrm{SO}(n)$ as follows.

Choose an orthonormal basis $(\epsilon_1, \ldots, \epsilon_{n+1})$ of $\mathbb{R}^{n+1}$. Then $\mathrm{SO}(n)$ can be looked at in the following way:

$$
\begin{aligned}
\mathrm{SO}(n) \simeq \mathfrak{K} \; &:= \; \{g \in \mathrm{SO}(n+1) \colon g(\epsilon_{n+1}) = \epsilon_{n+1}\} \\
&= \; \left\{ \begin{pmatrix} R & 0 \\ 0 & 1 \end{pmatrix} \colon R \in \mathrm{SO}(n) \right\} \subset \mathrm{SO}(n+1).
\end{aligned}
$$

The subgroup $\mathfrak{K}$ of $\mathrm{SO}(n+1)$ is also one of the two components of

$$
\{g \in \mathrm{SO}(n+1) \colon \mathrm{Ad}_P(g) := PgP^{-1} = g\},
$$

where

$$
P = \begin{pmatrix} 1_n & 0 \\ 0 & -1 \end{pmatrix}.
$$

An equivalence relation in $\mathrm{SO}(n+1)$ is defined by $g \mathcal{R} g' \Leftrightarrow g^{-1}g' \in \mathfrak{K}$ and the set of equivalence classes $\{[g] = g\mathfrak{K}/g \in \mathrm{SO}(n+1)\}$ is denoted $\mathrm{SO}(n+1)/\mathrm{SO}(n)$. Now it can easily be seen that

$$
\begin{aligned}
\mathrm{SO}(n+1)/\mathfrak{K} &\to S^n \\
[g] &\mapsto g(\epsilon_{n+1}).
\end{aligned}
$$

is a diffeomorphism.

Lifting a map into the sphere

Consider a map $u \colon \Omega \to S^n$, where Ω is an open, simply connected subset of $\mathbb{C}$. Then we can lift u to $\mathrm{SO}(n+1)$, i. e. we can find a map

$$
F \colon \Omega \to \mathrm{SO}(n+1),
$$

such that

$$
F(z)(\epsilon_{n+1}) = u(z).
$$

Write $F(z) = (e_1(z), \ldots e_n(z), u(z))$, then $e_1, \ldots, e_n$ form an orthonormal basis of $T_u S^n$. We can thus write

$$
du = \sum_{a=1}^{n} \phi^a e_a,
$$

where $\phi_a = \langle du, e_a \rangle$, and

$$de_a = \sum_{b=1}^{n} \omega_a^b e_b - \phi^a u, \quad u = 1, \ldots n,$$

where $\omega_b^a = \langle de_b, e_a \rangle$. The matrix

$$\alpha = \begin{pmatrix} 0 & \omega_2^1 & \cdots & \omega_n^1 & \phi^1 \\ \omega_1^2 & 0 & \cdots & \omega_n^2 & \phi^2 \\ \vdots & \vdots & \ddots & \vdots & \vdots \\ \omega_1^n & \omega_2^n & \cdots & 0 & \phi^n \\ -\phi^1 & -\phi^2 & \cdots & -\phi^n & 0 \end{pmatrix} \in T^*\Omega \otimes \mathrm{so}(n+1)$$

satisfies

$$dF = F\alpha. \tag{7.1}$$

We have seen in chapter 5, how the condition

$$d\alpha + \alpha \wedge \alpha = 0 \tag{7.2}$$

can be derived from (7.1). Furthermore, lemma 5.2 showed that if Ω is simply connected (which we assume here), and if (7.2) is valid, then there exists an $F : \Omega \to \mathrm{SO}(n+1)$, such that

$$\begin{cases} F(z_0) &= F_0, \\ dF &= F\alpha, \end{cases}$$

where $F_0 \in \mathrm{SO}(n+1)$ may be chosen arbitrarily. The passing from F to α can be looked upon as a kind of linearization. The non-linear condition $F \in \mathrm{SO}(n+1)$ is replaced be the linear one $\alpha + {}^t\alpha = 0$ (once we assume that $F_0 \in \mathrm{SO}(n+1)$).

We introduce the notation

$$[\alpha \wedge \beta] = \alpha \wedge \beta + \beta \wedge \alpha = [\beta \wedge \alpha]$$

for 1-forms $\alpha, \beta \in T^*\Omega \otimes \mathrm{gl}(n+1, \mathbb{C})$, where we write

$$\alpha \wedge \beta = \sum_{j=1}^{n} \begin{pmatrix} \alpha_j^1 \wedge \beta_1^j & \cdots & \alpha_j^1 \wedge \beta_n^j \\ \vdots & \ddots & \vdots \\ \alpha_j^n \wedge \beta_1^j & \cdots & \alpha_j^n \wedge \beta_n^j \end{pmatrix}$$

for

$$\alpha = \begin{pmatrix} \alpha_1^1 & \cdots & \alpha_n^1 \\ \vdots & \ddots & \vdots \\ \alpha_1^n & \cdots & \alpha_n^n \end{pmatrix}, \quad \beta = \begin{pmatrix} \beta_1^1 & \cdots & \beta_n^1 \\ \vdots & \ddots & \vdots \\ \beta_1^n & \cdots & \beta_n^n \end{pmatrix}.$$

This can be generalized to general Lie algebras by setting

$$[\alpha \wedge \beta](\xi, \eta) = [\alpha(\xi), \beta(\eta)] - [\alpha(\eta), \beta(\xi)].$$

Thus (7.2) may be rewritten

$$d\alpha + \frac{1}{2}[\alpha \wedge \alpha] = 0.$$

Decomposition of $so(n+1)$

We are going to decompose $so(n+1)$ and hence the α defined above. To this end, consider the operator

$$\begin{aligned} \mathrm{Ad}_P \colon so(n+1) &\rightarrow so(n+1) \\ \xi &\mapsto P\xi P^{-1}, \end{aligned}$$

which we have seen earlier. Since it is an involution, this operator acts linearly on $so(n+1)$ with eigenvalues ± 1. Moreover, it establishes a Lie algebra automorphism, i. e. we have

$$[\mathrm{Ad}_P(\xi), \mathrm{Ad}_P(\eta)] = \mathrm{Ad}_P([\xi, \eta]).$$

Now, let

$$\begin{aligned} so(n+1)_0 &= \{\xi \in so(n+1) \colon \mathrm{Ad}_P(\xi) = (-1)^0 \xi = \xi\}, \\ so(n+1)_1 &= \{\xi \in so(n+1) \colon \mathrm{Ad}_P(\xi) = (-1)^1 \xi = -\xi\}. \end{aligned}$$

Recall that we considered earlier the subgroup $\mathfrak{K}$ of $SO(n+1)$,

$$\mathfrak{K} = \left\{ \begin{pmatrix} R & 0 \\ 0 & 1 \end{pmatrix} \colon R \in SO(n) \right\} \simeq SO(n).$$

It turns out that $so(n+1)_0$ is the Lie algebra corresponding to $\mathfrak{K}$. Moreover, if $\xi \in so(n+1)_a, \eta \in so(n+1)_b$ then, using the fact that Ad_P is an automorphism, it can easily be checked that $[\xi, \eta] \in so(n+1)_c$, where $c \equiv a+b \pmod 2$. We now decompose α according to the decomposition $so(n+1) = so(n+1)_0 \oplus so(n+1)_1$:

$$\alpha = \alpha_0 + \alpha_1,$$

where

$$\alpha_0 = \begin{pmatrix} 0 & \omega_2^1 & \cdots & \omega_n^1 & 0 \\ \omega_1^2 & 0 & & \vdots & 0 \\ \vdots & & \ddots & \omega_n^{n-1} & \vdots \\ \omega_1^n & \cdots & \omega_{n-1}^n & 0 & 0 \\ 0 & 0 & \cdots & 0 & 0 \end{pmatrix} \in T^*\Omega \otimes so(n+1)_0,$$

and

$$
\alpha_1 = \begin{pmatrix} 0 & \cdots & 0 & \phi^1 \\ \vdots & \ddots & \vdots & \vdots \\ 0 & \cdots & 0 & \phi^n \\ -\phi^1 & \cdots & -\phi^n & 0 \end{pmatrix} \in T^*\Omega \otimes \mathrm{so}(n+1)_1.
$$

Then the integrability condition (7.2), written in the form

$$
d\alpha + \frac{1}{2}[\alpha \wedge \alpha] = 0,
$$

and projected onto $\mathrm{so}(n+1)_0$ and $\mathrm{so}(n+1)_1$, respectively, implies

$$
d\alpha_0 + \frac{1}{2}[\alpha_0 \wedge \alpha_0] + \frac{1}{2}[\alpha_1 \wedge \alpha_1] = 0 \tag{7.3}
$$

and

$$
d\alpha_1 + [\alpha_0 \wedge \alpha_1] = 0. \tag{7.4}
$$

Remark Note that (7.4) is a linear condition on α_1. This will be an important fact for what is to follow.

The harmonic map equation

We now assume that u is harmonic, that is

$$
\Delta u + u|du|^2 = 0,
$$

which is equivalent to

$$
d(*du) \parallel u. \tag{7.5}
$$

Recall that

$$
\begin{cases} du & = \displaystyle\sum_{a=1}^{n} \phi^a e_a, \\ de_a & = \displaystyle\sum_{b=1}^{n} \omega_a^b e_b - \phi^a u, \quad a = 1, \ldots, n. \end{cases}
$$

We obtain therefore

$$
d(*du) = \sum_{b=1}^{n} \left[d(*\phi^b) + \sum_{a=1}^{n} \omega_a^b \wedge (*\phi^a) \right] e_b - \sum_{a=1}^{n} \phi^a \wedge (*\phi^a) u.
$$

Thus (7.5) means

$$
d(*\phi^b) + \sum_{a=1}^{n} \omega_a^b \wedge (*\phi^a) = 0, \quad b = 1, \ldots, n,
$$

which may be written in the form

$$d(*\alpha_1) + [\alpha_0 \wedge *\alpha_1] = 0.$$

Forgetting about u for the moment and thus focusing on α, we see that we have to study the system

$$d\alpha_0 + \frac{1}{2}[\alpha_0 \wedge \alpha_0] + \frac{1}{2}[\alpha_1 \wedge \alpha_1] \quad = \quad 0, \qquad (7.6)$$

$$d\alpha_1 + [\alpha_0 \wedge \alpha_1] \quad = \quad 0, \qquad (7.7)$$

$$d(*\alpha_1) + [\alpha_0 \wedge (*\alpha_1)] \quad = \quad 0. \qquad (7.8)$$

We think of (7.7), (7.8) as a nonlinear Cauchy–Riemann system, in contrast to the linear one

$$\left\{ \begin{array}{rcl} d\beta & = & 0, \\ d(*\beta) & = & 0, \end{array} \right.$$

arising from a harmonic function $f \colon \Omega \to \mathbb{R}$ by setting $\beta = df$. The idea to treat (7.6)–(7.8) is to generalize the linear situation, in which case we are led to

$$d\beta_\lambda = 0$$

for

$$\beta_\lambda = \frac{\lambda + \lambda^{-1}}{2}\beta + \frac{\lambda - \lambda^{-1}}{2i}(*\beta) = \lambda^{-1}\beta' + \lambda\beta'', \quad \lambda \in \mathbb{C}^*,$$

where $\beta' = \beta\left(\frac{\partial}{\partial z}\right)dz$ and $\beta'' = \beta\left(\frac{\partial}{\partial \bar{z}}\right)d\bar{z}$. In the situation that we are dealing with now, we use

- the fact that (7.7) and (7.8) are the same equation in α_1 and $*\alpha_1$, respectively, and linear with respect to α_1,
- the property

$$[(\cos(\theta)\,\alpha_1 + \sin(\theta)\,(*\alpha_1)) \wedge (\cos(\theta)\,\alpha_1 + \sin(\theta)\,(*\alpha_1))] = [\alpha_1 \wedge \alpha_1]$$

for all $\theta \in \mathbb{C}$, which is easy to verify.

We then observe that (7.6)–(7.8) is equivalent to

$$d\alpha_\lambda + \frac{1}{2}[\alpha_\lambda \wedge \alpha_\lambda] = 0, \quad \lambda \in \mathbb{C}^*,$$

where

$$\alpha_\lambda = \lambda^{-1}\alpha_1' + \alpha_0 + \lambda\alpha_1'' = \frac{\lambda + \lambda^{-1}}{2}\alpha_1 + \alpha_0 + \frac{\lambda - \lambda^{-1}}{2i}(*\alpha_1).$$

This proves the following.

Theorem 7.1 *Assume that Ω is simply connected. Let $u: \Omega \to S^n$ be a map having the lift $F: \Omega \to \mathrm{SO}(n+1)$. Set $\alpha = F^{-1}.dF$, and decompose*

$$\alpha = \alpha_0 + \alpha_1$$

according to the decomposition

$$\mathrm{so}(n+1) = \mathrm{so}(n+1)_0 \oplus \mathrm{so}(n+1)_1.$$

Decompose further $\alpha_1 = \alpha'_1 + \alpha''_1$, where $\alpha'_1 = \alpha_1 \left(\frac{\partial}{\partial z}\right) dz$ and $\alpha''_1 = \alpha_1 \left(\frac{\partial}{\partial \bar{z}}\right) d\bar{z}$. Then u is harmonic if and only if the condition

$$d\alpha_\lambda + \frac{1}{2}[\alpha_\lambda \wedge \alpha_\lambda] = 0 \tag{7.9}$$

holds for all $\lambda \in \mathbb{C}^*$, where

$$\alpha_\lambda = \lambda^{-1}\alpha'_1 + \alpha_0 + \lambda\alpha''_1.$$

Conversely, each family of 1-forms α_λ on Ω of the form $\alpha_\lambda = \lambda^{-1}\alpha'_1 + \alpha_0 + \lambda\alpha''_1$, with coefficients satisfying the algebraic conditions above, and which is a solution of (7.9), gives rise to a S^1-family of harmonic maps by the following algorithm.

- Integrate

$$\begin{cases} F_\lambda(z_0) &= F_0 \\ dF_\lambda &= F_\lambda\alpha_\lambda \quad \text{on } \Omega \end{cases}$$

  for all $\lambda \in S^1$.
- Set $u_\lambda = [F_\lambda] = F_\lambda(\epsilon_{n+1})$.

Then, for any $\lambda \in S^1$, the map u_λ is harmonic into S^n. We omit the details.

Remark. In general, α_λ has coefficients in $\mathrm{so}(n+1) \otimes \mathbb{C}$. However, if $\lambda \in S^1 \subset \mathbb{C}^*$, then $\alpha_\lambda \in T^*\Omega \otimes \mathrm{so}(n+1)$. We thus recover the result of chapter 5 for harmonic maps into S^2, i. e. the existence of a *conjugate family* of harmonic maps.

7.2   Generalizations

We may replace $\mathrm{SO}(n+1)$ by any compact Lie group $\mathfrak{G}$ and $\mathfrak{K} \simeq \mathrm{SO}(n)$ by a subgroup of $\mathfrak{G}$. Suppose $\mathfrak{G}_\tau$ is defined from an automorphism $\tau: \mathfrak{G} \to \mathfrak{G}$ such that $\tau^2 = \mathbf{1}_\mathfrak{G}$ by

$$\mathfrak{G}_\tau = \{g \in G: \tau(g) = g\}$$

(τ plays the role of Ad_P), and let $(\mathfrak{G}_\tau)_0$ be the connected component of $\mathfrak{G}_\tau$ containing the identity. If $(\mathfrak{G}_\tau)_0 \subset \mathfrak{K} \subset \mathfrak{G}_\tau$, then the same results as before hold for harmonic maps $u \colon \Omega \to \mathfrak{G}/\mathfrak{K}$. More precisely, we proceed as follows.

First lift u to $\mathfrak{G}$, i. e. find a map $F \colon \Omega \to \mathfrak{G}$ such that u is the composition of F with the projection $\mathfrak{G} \to \mathfrak{G}/\mathfrak{K}$. Then set

$$\alpha = F^{-1}dF \in T^*\Omega \otimes \mathfrak{g},$$

where $\mathfrak{g}$ is the Lie algebra of $\mathfrak{G}$. We have the Cartan decomposition

$$\mathfrak{g} = \mathfrak{g}_0 \oplus \mathfrak{g}_1 =: \mathfrak{k} \oplus \mathfrak{g}_1,$$

where

$$\mathfrak{g}_a = \{\xi \in \mathfrak{g} \colon d\tau_1(\xi) = (-1)^a \xi\}, \quad a = 0, 1.$$

Note that $\mathfrak{k} = \mathfrak{g}_0$ is the Lie algebra of $\mathfrak{K}$. We split $\alpha = \alpha_0 + \alpha_1$ according to this decomposition. We let $\alpha_1' = \alpha_1 \left(\frac{\partial}{\partial z}\right) dz$ and $\alpha_1'' = \alpha_1 \left(\frac{\partial}{\partial \bar{z}}\right) d\bar{z}$, and we set

$$\alpha_\lambda = \lambda^{-1}\alpha_1' + \alpha_0 + \lambda \alpha_1''.$$

Then u is harmonic if and only if

$$d\alpha_\lambda + \frac{1}{2}[\alpha_\lambda \wedge \alpha_\lambda] = 0 \tag{7.10}$$

for all $\lambda \in \mathbb{C}^*$. Moreover, if (7.10) holds, then we get an S^1-family of harmonic maps by integrating

$$\begin{cases} F_\lambda(z_0) &= F_0, \\ dF_\lambda &= F_\lambda \alpha_\lambda, \end{cases}$$

and setting $u_\lambda = [F_\lambda]$.

An example

The sphere S^2 can also be represented by $\mathrm{SU}(2)/\mathrm{U}(1)$, where

$$\begin{aligned} \mathrm{SU}(2) &= \{g \in \mathrm{GL}(2, \mathbb{C}) \colon g^\dagger g = {}^t\bar{g}g = \mathbf{1},\ \det g = 1\} \\ &= \left\{\begin{pmatrix} a & -\bar{b} \\ b & \bar{a} \end{pmatrix} \colon a, b \in \mathbb{C},\ |a|^2 + |b|^2 = 1\right\} \simeq S^3, \end{aligned}$$

and

$$\mathrm{U}(1) = \{g \in \mathrm{SU}(2) \colon \mathrm{Ad}_P(g) = g\},$$

where

$$\mathrm{Ad}_P(g) = \begin{pmatrix} 1 & 0 \\ 0 & -1 \end{pmatrix} g \begin{pmatrix} 1 & 0 \\ 0 & -1 \end{pmatrix}.$$

Moreover, SU(2) acts on

$$\mathbb{R}^3 \simeq V \;\; := \;\; \{M \in \mathrm{M}(2, \mathbb{C}) \colon \mathrm{tr}\, M = 0, \; {}^t\bar{M} = M\}$$
$$= \left\{ \begin{pmatrix} z & x - iy \\ x + iy & -z \end{pmatrix} \colon x, y, z \in \mathbb{R} \right\}$$

isometrically by $M \mapsto gMg^{-1}$. Pick an $M \in V$ and a $g \in \mathrm{SU}(2)$. Then $\det M = -(x^2 + y^2 + z^2)$ (if M is represented by x, y, z as above). Since $\det(gMg^{-1}) = \det M$, the mapping $M \mapsto gMg^{-1}$ may be looked upon as an element of $\mathrm{SO}(3)$. More precisely, the mapping

$$\Phi \colon \mathrm{SU}(2) \;\; \to \;\; \mathrm{SO}(3)$$
$$g \;\; \mapsto \;\; [M \mapsto gM{}^t\bar{g}]$$

is a twofold covering (actually $\mathrm{SU}(2)$ is the spin group $\mathrm{Spin}(3)$).

Let now

$$\sigma_3 = \begin{pmatrix} 1 & 0 \\ 0 & -1 \end{pmatrix} \in V,$$

then

$$\mathrm{SU}(2) \;\; \to V$$
$$g \;\; \mapsto \;\; g\sigma_3 {}^t\bar{g}$$

maps $\mathrm{SU}(2)$ into $S^2 \subset V$. The fibers of this mapping - the Hopf fibration - are exactly the classes in $\mathrm{SU}(2)/\mathrm{U}(1)$. This gives an identification of $\mathrm{SU}(2)/\mathrm{U}(1)$ with S^2.

We are thus led to two different representations of a harmonic map $u \colon \Omega \to S^2$:

- by

$$\alpha_\lambda \;\; = \;\; \lambda^{-1} \begin{pmatrix} 0 & 0 & (\phi^1)' \\ 0 & 0 & (\phi^2)' \\ -(\phi^1)' & -(\phi^2)' & 0 \end{pmatrix} + \omega_1^2 \begin{pmatrix} 0 & -1 & 0 \\ 1 & 0 & 0 \\ 0 & 0 & 0 \end{pmatrix}$$
$$+ \lambda \begin{pmatrix} 0 & 0 & (\phi^1)'' \\ 0 & 0 & (\phi^2)'' \\ -(\phi^1)'' & -(\phi^2)'' & 0 \end{pmatrix} \in T^*\Omega \otimes \mathfrak{so}(3),$$

- and by

$$\alpha_\lambda \;\; = \;\; \frac{\lambda^{-1}}{2} \begin{pmatrix} 0 & (-\phi^1 + i\phi^2)' \\ (\phi^1 + i\phi^2)' & 0 \end{pmatrix} + \frac{\omega_1^2}{2} \begin{pmatrix} -i & 0 \\ 0 & i \end{pmatrix}$$
$$+ \frac{\lambda}{2} \begin{pmatrix} 0 & (-\phi^1 + i\phi^2)'' \\ (\phi^1 + i\phi^2)'' & 0 \end{pmatrix} \in T^*\Omega \otimes \mathfrak{su}(2).$$

7.3 A new setting: loop groups

We are going to take a different point of view by putting the parameter λ in the target. Let $\mathfrak{G}$, $\mathfrak{K}$, and τ be as in the previous section. We may keep in mind the case $\mathfrak{G} = \mathrm{SO}(n+1)$, $\mathfrak{K} = \mathrm{SO}(n)$, and $\tau = \mathrm{Ad}_P$. However, this works actually for any Lie group. Let's first define the loop groups:

$$\begin{aligned}
\Lambda\mathfrak{G} &= \{\lambda \mapsto g_\lambda : \lambda \in S^1,\ g_\lambda \in \mathfrak{G}\}, \\
\Lambda\mathfrak{G}^{\mathbb{C}} &= \{\lambda \mapsto g_\lambda : \lambda \in S^1,\ g_\lambda \in \mathfrak{G}^{\mathbb{C}}\},
\end{aligned}$$

where $\mathfrak{G}^{\mathbb{C}}$ is the complexification of $\mathfrak{G}$, e. g. for $\mathfrak{G} = \mathrm{SO}(n+1)$ or $\mathfrak{G} = \mathrm{U}(n)$,

$$\begin{aligned}
\mathrm{SO}(n+1)^{\mathbb{C}} &= \{g \in \mathrm{GL}(n+1, \mathbb{C}) : {}^t g g = \mathbf{1}_{n+1},\ \det g = 1\}, \\
\mathrm{U}(n)^{\mathbb{C}} &= \mathrm{GL}(n, \mathbb{C}).
\end{aligned}$$

We also need the twisted loop groups

$$\begin{aligned}
\Lambda\mathfrak{G}^{\mathbb{C}}_\tau &= \{\lambda \mapsto g_\lambda : \lambda \in S^1,\ g_\lambda \in \mathfrak{G}^{\mathbb{C}},\ \tau(g_\lambda) = g_{-\lambda}\}, \\
\Lambda\mathfrak{G}_\tau &= \Lambda\mathfrak{G}^{\mathbb{C}}_\tau \cap \Lambda\mathfrak{G}.
\end{aligned}$$

We endow these sets with a topology, for instance by using the Fourier decomposition

$$g_\lambda = \sum_{k \in \mathbb{Z}} \hat{g}_k \lambda^k,$$

and setting

$$\|g_\lambda\|^2_{H^s} = \sum_{k \in \mathbb{Z}} |\hat{g}_k|^2 (1 + k^2)^{s/2}$$

for $s > \frac{1}{2}$. Then the point-wise product

$$[\lambda \mapsto g_\lambda].[\lambda \mapsto h_\lambda] := [\lambda \mapsto g_\lambda h_\lambda]$$

is continuous and gives these sets some "Lie group" structure. The corresponding Lie algebras

$$\Lambda\mathfrak{g}^{\mathbb{C}} = \{\lambda \mapsto \xi_\lambda : \lambda \in S^1,\ \xi_\lambda \in \mathfrak{g}^{\mathbb{C}}\},$$

$\Lambda\mathfrak{g}$, $\Lambda\mathfrak{g}^{\mathbb{C}}_\tau$, etc have the Lie brackets

$$[[\lambda \mapsto \xi_\lambda], [\lambda \mapsto \eta_\lambda]] := [\lambda \mapsto [\xi_\lambda, \eta_\lambda]].$$

Note that the twisting condition corresponds to

$$\tau(\xi_\lambda) = \xi_{-\lambda}$$

for $[\lambda \mapsto \xi_\lambda] \in \Lambda\mathfrak{g}_\tau^{\mathbb{C}}$. But using Fourier's decomposition

$$\xi_\lambda = \sum_{k \in \mathbb{Z}} \hat{\xi}_k \lambda^k$$

once more, we see that this is equivalent to

$$\begin{cases} \hat{\xi}_{2k} & \in & \mathfrak{g}_0^{\mathbb{C}} \\ \hat{\xi}_{2k+1} & \in & \mathfrak{g}_1^{\mathbb{C}} \end{cases}$$

for all $k \in \mathbb{Z}$.

Link with harmonic maps

Suppose $u: \Omega \to \mathfrak{G}/\mathfrak{K}$ is a harmonic map. We have seen how to construct from it a family of 1-forms

$$\alpha_\lambda = \lambda^{-1}\alpha_1' + \alpha_0 + \lambda\alpha_1'',$$

which we may regard as a 1-form with coefficients in $\Lambda\mathfrak{g}^{\mathbb{C}}$. But furthermore, we have

- $\alpha_\lambda \in T^*\Omega \otimes \Lambda\mathfrak{g}$ for $\lambda \in S^1$ (reality condition), and
- $\alpha_1', \alpha_1'' \in T^*\Omega \otimes \mathfrak{g}_1^{\mathbb{C}}$, $\alpha_0 \in T^*\Omega \otimes \mathfrak{g}_0$. That is, $\alpha_\lambda \in T^*\Omega \otimes \Lambda\mathfrak{g}_\tau^{\mathbb{C}}$.

We have a simple characterization of harmonic maps in terms of loop groups.

We introduce further the loop groups

$$\begin{aligned} \Lambda^+\mathfrak{G}_\tau^{\mathbb{C}} &= \{[\lambda \mapsto g_\lambda] \in \Lambda\mathfrak{G}_\tau^{\mathbb{C}}: g_\lambda = \sum_{k \geq 0} \hat{g}_k \lambda^k\} \\ &= \{[\lambda \mapsto g_\lambda] \in \Lambda\mathfrak{G}_\tau^{\mathbb{C}}: g_\lambda \text{ admits a holomorphic} \\ &\qquad\qquad\qquad\qquad \text{extension inside the disk } |\lambda| \leq 1\}, \\ \Lambda^-\mathfrak{G}_\tau^{\mathbb{C}} &= \{[\lambda \mapsto g_\lambda] \in \Lambda\mathfrak{G}_\tau^{\mathbb{C}}: g_\lambda = \sum_{k \leq 0} \hat{g}_k \lambda^k\} \\ &= \{[\lambda \mapsto g_\lambda] \in \Lambda\mathfrak{G}_\tau^{\mathbb{C}}: [\lambda \mapsto g_{\lambda^{-1}}] \in \Lambda^+\mathfrak{G}_\tau^{\mathbb{C}}\} \\ \Lambda_\star^-\mathfrak{G}_\tau^{\mathbb{C}} &= \{[\lambda \mapsto g_\lambda] \in \Lambda^-\mathfrak{G}_\tau^{\mathbb{C}}: g_\infty = \mathbf{1}\}, \end{aligned}$$

and their corresponding Lie algebras $\Lambda^+\mathfrak{g}_\tau^{\mathbb{C}}$, etc. (Note that the loops in $\Lambda^-\mathfrak{G}_\tau^{\mathbb{C}}$ can be extended holomorphically to $(\mathbb{C} \cup \{\infty\}) \cap \{|\lambda| \geq 1\}$, so the condition $g_\infty = \mathbf{1}$ makes sense.) In particular, for any $[\lambda \mapsto \xi_\lambda] \in \Lambda\mathfrak{g}_\tau^{\mathbb{C}}$, the splitting

$$\xi_\lambda = \left(\sum_{k < 0} \hat{\xi}_k \lambda^k\right) + \left(\sum_{k \geq 0} \hat{\xi}_k \lambda^k\right)$$

shows that

$$\Lambda \mathfrak{g}_\tau^{\mathbb{C}} = \Lambda_\star^- \mathfrak{g}_\tau^{\mathbb{C}} \oplus \Lambda^+ \mathfrak{g}_\tau^{\mathbb{C}}.$$

We denote by $[\xi_\lambda]_{\Lambda_\star^- \mathfrak{g}_\tau^{\mathbb{C}}}$ the component of $\xi_\lambda \in \Lambda \mathfrak{g}_\tau^{\mathbb{C}}$ in $\Lambda_\star^- \mathfrak{g}_\tau^{\mathbb{C}}$, according to this decomposition. Then we have the following.

Proposition 7.1 *Let $F_\lambda \colon \Omega \to \Lambda \mathfrak{G}_\tau$ be a map such that*

$$[F_\lambda^{-1} dF_\lambda]_{\Lambda_\star^- \mathfrak{g}_\tau^{\mathbb{C}}} = \lambda^{-1}\beta, \tag{7.11}$$

where $\beta \in T^*\Omega \otimes \mathfrak{g}_1^{\mathbb{C}}$ satisfies

$$\beta\left(\frac{\partial}{\partial \bar{z}}\right) = 0. \tag{7.12}$$

Then, for any λ, the map $z \mapsto [F_\lambda](z)$ is harmonic.

Any map F_λ as in Proposition 7.1 will be named an *extended lift of a harmonic map*.
Proof. Consider the family of 1-forms

$$\alpha_\lambda = F_\lambda^{-1} dF_\lambda.$$

The twisting condition $F_\lambda \in \Lambda \mathfrak{G}_\tau$ implies that α_λ is also twisted, that is,

$$\alpha_\lambda = \sum_{k \in \mathbb{Z}} \hat{\alpha}_k \lambda^k$$

with $\hat{\alpha}_{2k} \in \mathfrak{g}_0^{\mathbb{C}}$ and $\hat{\alpha}_{2k+1} \in \mathfrak{g}_1^{\mathbb{C}}$. The condition (7.11) shows that $\hat{\alpha}_k = 0$ if $k < -1$. By the reality condition $F_\lambda \in \Lambda \mathfrak{G}$ we have also $\alpha_\lambda \in \Lambda \mathfrak{g}$ and hence

$$\sum_{k \geq -1} \overline{\hat{\alpha}_k} \lambda^{-k} = \sum_{k \geq -1} \hat{\alpha}_k \lambda^k$$

for $\lambda \in S^1$. Hence $\overline{\hat{\alpha}_{-k}} = \hat{\alpha}_k$. So we end up with

$$\alpha_\lambda = \lambda^{-1} \overline{\hat{\alpha}_1} + \hat{\alpha}_0 + \lambda \hat{\alpha}_1 = \lambda^{-1} \beta + \hat{\alpha}_0 + \lambda \bar{\beta},$$

with $\hat{\alpha}_0 \in \mathfrak{g}_0$. Since we have furthermore $\beta\left(\frac{\partial}{\partial \bar{z}}\right) = 0$ and hence $\bar{\beta}\left(\frac{\partial}{\partial z}\right) = 0$, this is exactly the decomposition as required in Theorem 7.1. However, in this case the condition (7.9) holds trivially because of the definition of α_λ. Therefore, Theorem 7.1 implies the claim. $\qquad \square$

Gauge transformations

The lift F of a harmonic map $u: \Omega \to \mathfrak{G}/\mathfrak{K}$ is of course arbitrary, since for any smooth map $h: \Omega \to \mathfrak{K}$, the map $Fh: \Omega \to \mathfrak{G}$ lifts the same u.

Suppose that a lift F of u and a smooth map $h: \Omega \to \mathfrak{K}$ are given, and let $\tilde{F} = Fh$. Assume for simplicity that $F_\lambda(z_0) = \mathbf{1} = h(z_0)$. We want to compute the corresponding $\tilde{F}_\lambda$. To this end, first compute

$$\tilde{\alpha} = \tilde{F}^{-1} d\tilde{F} = h^{-1} F^{-1}(dF h + F dh) = h^{-1} \alpha h + h^{-1} dh = \tilde{\alpha}_0 + \tilde{\alpha}_1,$$

where $\tilde{\alpha}_0 = h^{-1} \alpha_0 h + h^{-1} dh$ and $\tilde{\alpha}_1 = h^{-1} \alpha_1 h$, for the usual splitting $\alpha = \alpha_0 + \alpha_1$. Hence we have

$$\tilde{\alpha}_\lambda = \lambda^{-1}(h^{-1} \alpha_1' h) + (h^{-1} \alpha_0 h + h^{-1} dh) + \lambda(h^{-1} \alpha_1'' h),$$

which means

$$\tilde{F}_\lambda^{-1} d\tilde{F}_\lambda = \tilde{\alpha}_\lambda = h^{-1} \alpha_\lambda h + h^{-1} dh = (F_\lambda h)^{-1} d(F_\lambda h),$$

and $\tilde{F}_\lambda(z_0) = F_\lambda(z_0)h(z_0) = \mathbf{1}$. So we see that

$$\tilde{F}_\lambda = F_\lambda h.$$

Thus harmonic maps $u: \Omega \to \mathfrak{G}/\mathfrak{K}$ correspond bijectively to classes of maps

$$[F_\lambda] \in \Lambda \mathfrak{G}_\tau / \mathfrak{K},$$

such that $[F_\lambda^{-1} dF_\lambda]_{\Lambda_*^- \mathfrak{G}_\tau^{\mathbb{C}}} = \lambda^{-1} \beta$ and $\beta\left(\frac{\partial}{\partial \bar{z}}\right) = 0$.

Twistor interpretation

Comparing the results with those of the previous chapter, we see some similarities. We may therefore give the following interpretation.

- $Z = \Lambda \mathfrak{G}_\tau / \mathfrak{K}$ is a kind of twistor space.
- The condition $[F_\lambda^{-1} dF_\lambda]_{\Lambda_*^- \mathfrak{G}_\tau^{\mathbb{C}}} = \lambda^{-1} \beta$ is a horizontality condition.
- The condition $[F_\lambda^{-1} dF_\lambda]_{\Lambda_*^- \mathfrak{G}_\tau^{\mathbb{C}}}\left(\frac{\partial}{\partial \bar{z}}\right) = 0$ means that F_λ is holomorphic in a certain way.

7.4   Examples

i) Let $u: \Omega \to SO(3)/SO(2) \simeq S^2$ be conformal. Assume for instance that u is even holomorphic, i. e.

$$\begin{cases} u \times \dfrac{\partial u}{\partial x} = \dfrac{\partial u}{\partial y}, \\ u \times \dfrac{\partial u}{\partial y} = -\dfrac{\partial u}{\partial x}, \end{cases}$$

or, equivalently,

$$*(u \times du) = du. \tag{7.13}$$

Let $F = (e_1, e_2, u)$ be a lift (or moving frame) belonging to u, and let

$$\alpha = F^{-1}dF = \begin{pmatrix} 0 & \omega_2^1 & \phi^1 \\ \omega_1^2 & 0 & \phi^2 \\ -\phi^1 & -\phi^2 & 0 \end{pmatrix} = \begin{pmatrix} 0 & \alpha_2^1 & \alpha_3^1 \\ \alpha_1^2 & 0 & \alpha_3^2 \\ \alpha_1^3 & \alpha_2^3 & 0 \end{pmatrix}.$$

We have by definition

$$du = e_1\phi^1 + e_2\phi^2,$$

and hence

$$u \times du = e_2\phi^1 - e_1\phi^2.$$

So (7.13) is equivalent to

$$\begin{cases} *\phi^1 & = & \phi^2, \\ *\phi^2 & = & -\phi^1, \end{cases}$$

and thus to

$$\begin{cases} (\phi^2)' & = & -i(\phi^1)', \\ (\phi^2)'' & = & i(\phi^1)''. \end{cases}$$

We find thus the representations

$$\alpha_1' = \begin{pmatrix} 0 & 0 & (\phi^1)' \\ 0 & 0 & (\phi^2)' \\ -(\phi^1)' & -(\phi^2)' & 0 \end{pmatrix} = (\phi^1)' \begin{pmatrix} 0 & 0 & 1 \\ 0 & 0 & -i \\ -1 & i & 0 \end{pmatrix},$$

and

$$\alpha_1'' = (\phi^1)'' \begin{pmatrix} 0 & 0 & 1 \\ 0 & 0 & i \\ -1 & -i & 0 \end{pmatrix}.$$

Set

$$A_0 = \begin{pmatrix} 0 & -1 & 0 \\ 1 & 0 & 0 \\ 0 & 0 & 0 \end{pmatrix}, \ A_+ = \begin{pmatrix} 0 & 0 & 1 \\ 0 & 0 & -i \\ -1 & i & 0 \end{pmatrix}, \ A_- = \begin{pmatrix} 0 & 0 & 1 \\ 0 & 0 & i \\ -1 & -i & 0 \end{pmatrix}.$$

Then

$$\alpha = (\phi^1)'A_+ + \omega_1^2 A_0 + (\phi^1)''A_-.$$

We let

$$u = \frac{1}{1 + |v|^2} \begin{pmatrix} v + \bar{v} \\ -i(v - \bar{v}) \\ 1 - |v|^2 \end{pmatrix},$$

where $v \colon \Omega \to \mathbb{C} \cup \{\infty\}$ is meromorphic, and we choose F such that

$$e_1 + ie_2 = \frac{i}{1 + |v|^2} \begin{pmatrix} i(v^2 - 1) \\ v^2 + 1 \\ 2iv \end{pmatrix}.$$

We then compute

$$d'u \quad := \quad \frac{\partial u}{\partial z}\, dz = \frac{\frac{\partial v}{\partial z}\, dz}{1 + |v|^2}(e_1 - ie_2) = (\phi^1)'(e_1 - ie_2),$$

$$d''u \quad := \quad \frac{\partial u}{\partial \bar{z}}\, d\bar{z} = \frac{\frac{\partial \bar{v}}{\partial \bar{z}}\, d\bar{z}}{1 + |v|^2}(e_1 + ie_2) = (\phi^1)''(e_1 + ie_2),$$

and

$$\omega_1^2 = \langle de_1, e_2 \rangle = \frac{i}{1 + |v|^2}(\bar{v}dv - vd\bar{v}) = -*d(\log(1 + |v|^2)).$$

Hence we see that

$$
F \quad = \quad \frac{1}{1 + |v|^2} \begin{pmatrix} 1 - \frac{v^2 + \bar{v}^2}{2} & i\frac{v^2 - \bar{v}^2}{2} & v + \bar{v} \\ i\frac{v^2 - \bar{v}^2}{2} & 1 + \frac{v^2 + \bar{v}^2}{2} & -i(v - \bar{v}) \\ -(v + \bar{v}) & i(v - \bar{v}) & 1 - |v|^2 \end{pmatrix}
$$

$$
= \quad \frac{1}{1 + |v|^2} \left[\frac{v^2}{2} A_+^2 + vA_+ + \mathbf{1} + \bar{v}A_- + \frac{\bar{v}}{2} A_-^2 \right]
$$

and

$$\alpha = \frac{d'v}{1 + |v|^2} A_+ + \frac{i}{1 + |v|^2}(\bar{v}dv - vd\bar{v})A_0 + \frac{d''v}{1 + |v|^2} A_-.$$

Thus we may construct F_λ from F directly by the substitution of v by $\lambda^{-1}v$ to obtain

$$F_\lambda = \frac{1}{1 + |v|^2} \left[\frac{\lambda^{-2}v^2}{2} A_+^2 + \lambda^{-1}vA_+ + \mathbf{1} + \lambda\bar{v}A_- + \frac{\lambda^2\bar{v}}{2} A_-^2 \right].$$

Note that u is thereby deformed according to

$$u_\lambda = \begin{pmatrix} \frac{\lambda^{-1} + \lambda}{2} & -\frac{\lambda^{-1} - \lambda}{2i} & 0 \\ \frac{\lambda^{-1} - \lambda}{2i} & \frac{\lambda^{-1} + \lambda}{2} & 0 \\ 0 & 0 & 1 \end{pmatrix} u,$$

hence the action of S^1 on u is just a rotation.

ii) We are looking for "rotationally symmetric" harmonic maps $u \colon \mathbb{C} \to S^2$ of the form

$$u(z) = u(x, y) = \begin{pmatrix} \sin \beta(x) \cos(y/a) \\ \sin \beta(x) \sin(y/a) \\ \cos \beta(x) \end{pmatrix} = u(x, y + 2\pi a)$$

for some function $\beta \colon \mathbb{R} \to \mathbb{R}$.

We then have

$$\frac{\partial u}{\partial x} = \beta'(x) e_1, \quad \frac{\partial u}{\partial y} = \frac{\sin \beta(x)}{a} e_2,$$

where

$$e_1 = \begin{pmatrix} \cos \beta(x) \cos(y/a) \\ \cos \beta(x) \sin(y/a) \\ -\sin \beta(x) \end{pmatrix}, \quad e_2 = \begin{pmatrix} -\sin(y/a) \\ \cos(y/a) \\ 0 \end{pmatrix}.$$

We compute further

$$\frac{\partial}{\partial x} \left(u \times \frac{\partial u}{\partial x} \right) + \frac{\partial}{\partial y} \left(u \times \frac{\partial u}{\partial y} \right) = \left(\beta''(x) - \frac{\sin \beta(x) \cos \beta(x)}{a^2} \right) e_2.$$

So u is harmonic if and only if β is a solution of the pendulum equation

$$\beta'' - \frac{\sin \beta \cos \beta}{a^2} = 0, \tag{7.14}$$

in which case it has conserved energy

$$(\beta')^2 - \frac{\sin^2 \beta}{a^2} = E_0. \tag{7.15}$$

Note that the Hopf differential is then equal to $E_0(dz)^2$. If E_0 happens to be positive, then we may—by a suitable rescaling $z \mapsto rz$ and $a \mapsto \frac{a}{r}$—assume that $E_0 = 1$. Similarly, the case $E_0 < 0$ can be reduced to $E_0 = -1$. Let $F = (e_1, e_2, u)$. Then we have $F = e^{\frac{y}{a} A_3} e^{\beta A_2}$ and $dF = F\alpha$, with

$$\alpha = \begin{pmatrix} 0 & -\frac{\cos \beta}{a} dy & \beta' dx \\ \frac{\cos \beta}{a} dy & 0 & \frac{\sin \beta}{a} dy \\ -\beta' dx & -\frac{\sin \beta}{a} dy & 0 \end{pmatrix}$$

$$= \frac{\cos \beta}{a} A_3 \, dy - \frac{\sin \beta}{a} A_1 \, dy + \beta' A_2 \, dx,$$

where

$$A_1 = \begin{pmatrix} 0 & 0 & 0 \\ 0 & 0 & -1 \\ 0 & 1 & 0 \end{pmatrix}, \quad A_2 = \begin{pmatrix} 0 & 0 & 1 \\ 0 & 0 & 0 \\ -1 & 0 & 0 \end{pmatrix}, \quad A_3 = \begin{pmatrix} 0 & -1 & 0 \\ 1 & 0 & 0 \\ 0 & 0 & 0 \end{pmatrix}.$$

(Note that $[A_a, A_b] = A_c$ with $c \equiv a + b \pmod 3$.)

A solution of the pendulum equation can be obtained by integrating ordinary differential equations, or, geometrically speaking, by integrating vector fields. One way to do this is to eliminate β by letting

$$U = \frac{\sin \beta}{2a}, \quad V = \frac{\beta'}{2}, \quad Z = -\frac{\cos \beta}{a}, \tag{7.16}$$

so that

$$\frac{d}{dx} \begin{pmatrix} U \\ V \\ Z \end{pmatrix} = \begin{pmatrix} -VZ \\ -UZ \\ 4UV \end{pmatrix}. \tag{7.17}$$

Exercise 1 Prove that (7.17) is equivalent to the assertion: There are a number $a \in \mathbb{R}$ and a function $\beta \colon \mathbb{R} \to \mathbb{R}$, such that (7.16) and the pendulum equation (7.14) hold. (Hint: use $4U^2 + Z^2 = \mathrm{const.}$)

Exercise 2 From the energy conservation (7.15) deduce that

$$dx = -\frac{dU}{\sqrt{\left(\frac{E_0}{4} + U^2\right)\left(\frac{1}{a^2} - 4U^2\right)}}.$$

Deduce the solution in terms of elliptic integrals and theta functions (see [32], [19]).

Now we are going to consider the extended lift F_λ of u, defined by

$$dF_\lambda = F_\lambda \alpha_\lambda,$$

and

$$
\begin{aligned}
\alpha_\lambda &= \lambda^{-1} \left[\frac{i \sin \beta}{2a} A_1 + \frac{\beta'}{2} A_2 \right] dz + \frac{\cos \beta}{a} A_3 \, dy \\
&\quad + \lambda \left[-\frac{i \sin \beta}{2a} A_1 + \frac{\beta'}{2} A_2 \right] d\bar{z} \\
&= \lambda^{-1}(iUA_1 + VA_2) \, dz - ZA_3 \, dy + \lambda(-iUA_1 + VA_2) \, d\bar{z}.
\end{aligned}
$$

This is equivalent to replacing (U, V, Z) by $(\lambda^{-1}U, \lambda^{-1}V, Z)$. Note that still

$$\frac{d}{dx} \begin{pmatrix} \lambda^{-1}U \\ \lambda^{-1}V \\ Z \end{pmatrix} = \begin{pmatrix} -(\lambda^{-1}V)Z \\ -(\lambda^{-1}U)Z \\ 2(\overline{(\lambda^{-1}U)}(\lambda^{-1}V) + (\lambda^{-1}U)\overline{(\lambda^{-1}V)}) \end{pmatrix}$$

for all $\lambda \in S^1$. Thus this one-parameter family of vector fields is embedded in the set of solutions $U, V: \mathbb{R} \to \mathbb{C}$, $Z: \mathbb{R} \to \mathbb{R}$ of

$$\frac{d}{dx} \begin{pmatrix} U \\ V \\ Z \end{pmatrix} = \begin{pmatrix} -VZ \\ -UZ \\ 2(\overline{U}V + U\overline{V}) \end{pmatrix}. \tag{7.18}$$

One may check directly that any solution of (7.18) leads to some

$$\alpha_\lambda = \lambda^{-1}(iUA_1 + VA_2) \, dz - ZA_3 \, dy + \lambda(-i\overline{U}A_1 + \overline{V}A_2) \, d\bar{z}$$

that solves

$$d\alpha_\lambda + \frac{1}{2}[\alpha_\lambda \wedge \alpha_\lambda] = 0.$$

Moreover, the following holds.

Proposition 7.2 *Any solution $U, V: \mathbb{R} \to \mathbb{C}$, $Z: \mathbb{R} \to \mathbb{R}$ of (7.18) is also a solution of*

$$d\eta_\lambda = [\eta_\lambda, \alpha_\lambda],$$

where

$$\eta_\lambda = \lambda^{-1}(UA_1 - iVA_2) + ZA_3 + \lambda(\overline{U}A_1 + i\overline{V}A_2),$$

and conversely.

Proof. We have

$$d\eta_\lambda = \lambda^{-1}(dU A_1 - idV A_2) + dZ A_3 + \lambda(d\overline{U}A_1 + id\overline{V}A_2)$$

and

$$\begin{aligned}[\eta_\lambda, \alpha_\lambda] &= \lambda^{-1}(-VZA_1 + iUZA_2) \, dx + 2(\overline{U}V + \overline{V}U)A_3 \, dx \\ &\quad + \lambda(-\overline{V}ZA_1 - i\overline{U}ZA_2) \, dx.\end{aligned}$$

So $d\eta_\lambda = [\eta_\lambda, \alpha_\lambda]$ holds if and only if

$$dU = -VZ \, dx, \quad dV = -UZ \, dx, \quad dZ = 2(\overline{U}V + \overline{V}U) \, dx,$$

which concludes the proof. $\qquad\square$

Another key observation is that

$$\begin{aligned}\alpha_\lambda &= i\left[\lambda^{-1}(UA_1 - iVA_2) + \frac{Z}{2}A_3\right] dz - i\left[\frac{Z}{2}A_3 + \lambda(\overline{U}A_1 + i\overline{V}A_2)\right] d\bar{z} \\ &= i\eta_\lambda \, dz - \beta_\lambda,\end{aligned}$$

where
$$\beta_\lambda = iZ A_3\, dx + 2i\lambda(\overline{U} A_1 + i\overline{V} A_2)\, dx.$$

This decomposition
$$i\eta_\lambda\, dz = \alpha_\lambda + \beta_\lambda$$

corresponds to
$$\Lambda so(3)_\tau^{\mathbb{C}} = \Lambda so(3)_\tau \oplus \Lambda_{\mathfrak{b}}^+ so(3)_\tau^{\mathbb{C}},$$

where
$$\Lambda_{\mathfrak{b}}^+ so(3)_\tau^{\mathbb{C}} = \{[\lambda \mapsto \xi_\lambda] \in \Lambda^+ so(3)_\tau^{\mathbb{C}} : \xi_0 = \hat{\xi}_0 \in \mathfrak{b}\}$$

and $\mathfrak{b} = \{it A_3 : t \in \mathbb{R}\}$ is the Lie algebra of the subgroup

$$\mathfrak{B} = \left\{ \begin{pmatrix} \cosh t & -i\sinh t & 0 \\ i\sinh t & \cosh t & 0 \\ 0 & 0 & 1 \end{pmatrix} : t \in \mathbb{R} \right\}$$

of $\mathfrak{K}^{\mathbb{C}} \simeq SO(2)^{\mathbb{C}}$. Hence, if

$$r\colon \Lambda so(3)_\tau \oplus \Lambda_{\mathfrak{b}}^+ so(3)_\tau^{\mathbb{C}} \to \Lambda so(3)_\tau$$

is the corresponding projection, then we have

$$d\eta_\lambda = [\eta_\lambda, r(i\eta_\lambda\, dz)].$$

This example shows that some harmonic maps can be obtained by integrating vector fields. We will see more about this in the next chapter.

All these observations have an interpretation in terms of some natural loop groups operations, this will be the subject of the next chapter.

Remark The introduction of a complex parameter goes back to K. Pohlmeyer in 1976 [69]. The first extensive use of these ideas is due to K. Uhlenbeck in 1989 [82] and N. Hitchin in 1990 [54] and was followed by many other works [11], [12], [21], [22], [30], [41], [47], [49], [51]. The idea of loop groups, originated by works by M. Sato [74], was developed by G. Segal and G. Wilson in 1985 [75]. An algebraic theory of loop groups is constructed in the book of A. Pressley and G. Segal (1986) [70].

8 Construction of finite type solutions

At the end of the previous chapter, we have seen a kind of harmonic map that is constructed from a solution of the differential equation

$$
\frac{d}{dx} \begin{pmatrix} U \\ V \\ Z \end{pmatrix} = \begin{pmatrix} -VZ \\ -UZ \\ 2(\overline{U}V + U\overline{V}) \end{pmatrix}.
$$

We want to find a similar way to construct a greater variety of harmonic maps. To this end, we need a pair of appropriate vector fields X_1, X_2 on some finite dimensional vector space V, such that $[X_1, X_2] = 0$. This allows us to integrate them, i. e. to find a $\eta \colon \mathbb{R}^2 \to V$, such that

$$
\frac{\partial \eta}{\partial x} = X_1 \circ \eta, \quad \frac{\partial \eta}{\partial y} = X_2 \circ \eta.
$$

Via some linear mapping

$$
\Pi \colon V \to T^*\Omega \otimes \Lambda \mathfrak{g}_\tau^{\mathbb{C}},
$$

for an appropriate Lie algebra $\mathfrak{g}$, we hope to find some solution $\alpha_\lambda = \Pi(\eta)$ of

$$
d\alpha_\lambda + \frac{1}{2}[\alpha_\lambda \wedge \alpha_\lambda] = 0,
$$

and thus a family of harmonic maps. This is done by the so called Adler-Kostant-Symes theory [4], [60], [77]. Our exposition can be compared with the one of F. Burstall and F. Pedit in [22].

8.1 Preliminary: the Iwasawa decomposition (for $\mathfrak{K}^{\mathbb{C}}$)

Let $\mathfrak{G}$, $\mathfrak{K} \subset \mathfrak{G}$ be Lie groups as before. Let $\mathfrak{K}^{\mathbb{C}}$ be the complexification of $\mathfrak{K}$. For instance,

$$
\begin{aligned}
\mathrm{SO}(n)^{\mathbb{C}} &= \{ M \in \mathrm{GL}(n, \mathbb{C}) \colon {}^t M M = \mathbf{1}, \ \det M = 1 \}, \\
\mathrm{U}(n)^{\mathbb{C}} &= \mathrm{GL}(n, \mathbb{C}), \\
\mathrm{SU}(n)^{\mathbb{C}} &= \mathrm{SL}(n, \mathbb{C}).
\end{aligned}
$$

We want to find a subgroup $\mathfrak{B}_{\mathfrak{K}} \subset \mathfrak{K}^{\mathbb{C}}$, such that for all $g \in \mathfrak{K}^{\mathbb{C}}$ there exists a unique decomposition $g = fb$ with $f \in \mathfrak{K}$, $b \in \mathfrak{B}_{\mathfrak{K}}$; $\mathfrak{B}_{\mathfrak{K}}$ is then called a *Borel* subgroup of $\mathfrak{K}^{\mathbb{C}}$. This is a generalization of matrix decompositions like $M = RT$ for $M \in \mathrm{GL}(n, \mathbb{R})$, $R \in \mathrm{SO}(n)$, and a triangular matrix T.

Example 1 Let $\mathfrak{K} = \mathrm{SO}(2)$. Then we may choose

$$\mathfrak{B}_{\mathrm{SO}(2)} = \left\{ \begin{pmatrix} \cos it & -\sin it \\ \sin it & \cos it \end{pmatrix} : t \in \mathbb{R} \right\} \subset \mathrm{SO}(2)^{\mathbb{C}}.$$

Since any $g \in \mathrm{SO}(2)^{\mathbb{C}}$ may be written in the form

$$g = \exp\left((\theta + it) \begin{pmatrix} 0 & -1 \\ 1 & 0 \end{pmatrix} \right), \quad \theta, t \in \mathbb{R},$$

we have indeed $g = Rb$ with

$$R = \exp\left(\theta \begin{pmatrix} 0 & -1 \\ 1 & 0 \end{pmatrix} \right) \in \mathrm{SO}(2), \quad b = \exp\left(it \begin{pmatrix} 0 & -1 \\ 1 & 0 \end{pmatrix} \right) \in \mathfrak{B}_{\mathrm{SO}(2)},$$

and the decomposition is unique. The Lie algebra of $\mathfrak{B}_{\mathfrak{K}}$ is in this case

$$\mathfrak{b}_{\mathrm{SO}(2)} = \left\{ t \begin{pmatrix} 0 & -i \\ i & 0 \end{pmatrix} : t \in \mathbb{R} \right\}.$$

Example 2 Suppose $\mathfrak{K} = \mathrm{SO}(3)$, let

$$X = \begin{pmatrix} 1 \\ -i \\ 0 \end{pmatrix} \in \mathbb{C},$$

and define

$$\mathfrak{B}_{\mathrm{SO}(3)} = \{ b \in \mathrm{SO}(3)^{\mathbb{C}} : bX = \nu X \text{ for a } \nu \in (0, \infty) \}.$$

Now pick a $g \in \mathrm{SO}(3)^{\mathbb{C}}$, and let $Y = gX$. Since X is isotropic (i. e. $X \neq 0$ and $(X, X) = 0$), the same holds for Y. This means that Y can be written in the form $Y = Y_1 - iY_2$ for real vectors $Y_1, Y_2 \neq 0$, where $|Y_1| = |Y_2|$ and $\langle Y_1, Y_2 \rangle = 0$. Hence there exist a unique $R \in \mathrm{SO}(3)$ and a unique number $\nu \in (0, \infty)$, such that

$$Y_1 = \nu R \begin{pmatrix} 1 \\ 0 \\ 0 \end{pmatrix}, \quad Y_2 = \nu R \begin{pmatrix} 0 \\ 1 \\ 0 \end{pmatrix}.$$

Set $b = R^{-1}g$, then we have $bX = \nu X$, i. e. $b \in \mathfrak{B}_{\mathrm{SO}(3)}$. So we have the unique decomposition $g = Rb$.

The Lie algebra of $\mathfrak{B}_{\mathrm{SO}(3)}$ is

$$\mathfrak{b}_{\mathrm{so}(3)} := \left\{ \begin{pmatrix} 0 & -it & -ix \\ it & 0 & x \\ ix & -x & 0 \end{pmatrix} : t \in \mathbb{R}, x \in \mathbb{C} \right\}.$$

Note that here $\mathfrak{B}_{\mathrm{SO}(3)}$ is not unique, a different choice of X (pick up any other isotropic nonvanishing vector of $\mathbb{C}^3$) would lead to another Borel subgroup.

Example 3 Let $\mathfrak{K} = \mathrm{U}(n)$, then it can be verified, that

$$\mathfrak{B}_{\mathrm{U}(n)} = \left\{ \begin{pmatrix} T_1^1 & & 0 \\ \vdots & \ddots & \\ T_1^n & \cdots & T_n^n \end{pmatrix} : T_1^1, \ldots, T_n^n \in (0, \infty),\ T_i^j \in \mathbb{C} \text{ for } i < j \right\}$$

works.

In all cases we have furthermore that the mapping

$$\begin{aligned} \mathfrak{K} \times \mathfrak{B}_{\mathfrak{K}} &\rightarrow \mathfrak{K}^{\mathbb{C}} \\ (f, b) &\mapsto fb \end{aligned}$$

is a diffeomorphism, a property that we shall summarize by the relation $\mathfrak{K}^{\mathbb{C}} = \mathfrak{K}.\mathfrak{B}_{\mathfrak{K}}$. On the level of Lie algebras, this means

$$\mathfrak{k} \otimes \mathbb{C} = \mathfrak{k} \oplus \mathfrak{b}_{\mathfrak{K}},$$

where $\mathfrak{k}$ and $\mathfrak{b}_{\mathfrak{K}}$ are the Lie algebras belonging to $\mathfrak{K}$ and $\mathfrak{B}_{\mathfrak{K}}$, respectively. We assume from now on that such a decomposition of $\mathfrak{K}^{\mathbb{C}}$ exists.

8.2 Application to loop Lie algebras

Consider again the Lie algebras

$$\begin{aligned} \Lambda \mathfrak{g}^{\mathbb{C}} &= \{[\lambda \mapsto \xi_\lambda] : \xi_\lambda \in \mathfrak{g} \otimes \mathbb{C}\}, \\ \Lambda \mathfrak{g} &= \{[\lambda \mapsto \xi_\lambda] : \xi_\lambda \in \mathfrak{g}\}, \\ \Lambda^+ \mathfrak{g}^{\mathbb{C}} &= \{[\lambda \mapsto \xi_\lambda] \in \Lambda \mathfrak{g}^{\mathbb{C}} : \xi_\lambda \text{ extends holomorphically to the unit disk}\}, \end{aligned}$$

and the corresponding twisted Lie algebras. Define furthermore

$$\Lambda^+_{\mathfrak{b}_{\mathfrak{K}}} \mathfrak{g}^{\mathbb{C}}_\tau = \{[\lambda \mapsto \xi_\lambda] \in \Lambda^+ \mathfrak{g}^{\mathbb{C}}_\tau : \hat{\xi}_0 \in \mathfrak{b}_{\mathfrak{K}}\},$$

with the usual notation

$$\xi_\lambda = \sum_{k \in \mathbb{Z}} \hat{\xi}_k \lambda^k.$$

The following decomposition

$$\sum_{k \in \mathbb{Z}} \hat{\xi}_k \lambda^k = \left(\sum_{k < 0} \hat{\xi}_k \lambda^k + (\hat{\xi}_0)_{\mathfrak{k}} + \sum_{k > 0} \overline{\hat{\xi}_{-k}} \lambda^k \right) + \left((\hat{\xi}_0)_{\mathfrak{b}_{\mathfrak{K}}} + \sum_{k > 0} \left(\hat{\xi}_k - \overline{\hat{\xi}_{-k}} \right) \lambda^k \right),$$

where $\hat{\xi}_0 = (\hat{\xi}_0)_{\mathfrak{k}} + (\hat{\xi}_0)_{\mathfrak{b}_{\mathfrak{R}}}$ is the splitting according to $\mathfrak{k} \otimes \mathbb{C} = \mathfrak{k} \oplus \mathfrak{b}_{\mathfrak{R}}$, shows that we have the Lie algebra splitting

$$\Lambda \mathfrak{g}_\tau^{\mathbb{C}} = \Lambda \mathfrak{g}_\tau \oplus \Lambda_{\mathfrak{b}_{\mathfrak{R}}}^+ \mathfrak{g}_\tau^{\mathbb{C}}. \tag{8.1}$$

Let

$$r : \Lambda \mathfrak{g}_\tau^{\mathbb{C}} \to \Lambda \mathfrak{g}_\tau \qquad \text{and} \qquad j : \Lambda \mathfrak{g}_\tau^{\mathbb{C}} \to \Lambda_{\mathfrak{b}_{\mathfrak{R}}}^+ \mathfrak{g}_\tau^{\mathbb{C}}$$

be the two accompanying projections.

Remark $\Lambda \mathfrak{g}_\tau$ is the Lie algebra of $\Lambda \mathfrak{G}_\tau$. One can also check easily that $\Lambda_{\mathfrak{b}_{\mathfrak{R}}}^+ \mathfrak{g}_\tau^{\mathbb{C}}$ is a loop Lie algebra: we shall see in paragraph 8.4 the associated loop Lie group. This will turn to be crucial in the Chapter 11.

8.3   The algorithm

Now we are ready to construct harmonic maps. First choose an odd number $d \in \mathbb{N}^*$, and set

$$\Lambda^d \mathfrak{g}_\tau = \left\{ [\lambda \mapsto \xi_\lambda] \in \Lambda \mathfrak{g}_\tau : \xi_\lambda = \sum_{k=-d}^{d} \hat{\xi}_k \lambda^k \right\}.$$

This is no longer a Lie algebra. However, it is a finite dimensional vector space. Define the two vector fields $X_1, X_2 : \Lambda^d \mathfrak{g}_\tau \to \Lambda \mathfrak{g}$ by

$$
\begin{aligned}
X_1(\xi_\lambda) &= [\xi_\lambda, r(\lambda^{d-1}\xi_\lambda)] &= -[\xi_\lambda, j(\lambda^{d-1}\xi_\lambda)], \\
X_2(\xi_\lambda) &= [\xi_\lambda, r(i\lambda^{d-1}\xi_\lambda)] &= -[\xi_\lambda, j(i\lambda^{d-1}\xi_\lambda)].
\end{aligned}
\tag{8.2}
$$

Note that

$$
\begin{cases}
r(\lambda^{d-1}\xi_\lambda) &= \lambda^{-1}\hat{\xi}_{-d} + (\hat{\xi}_{1-d})_{\mathfrak{k}} + \lambda\overline{\hat{\xi}_{-d}}, \\
r(i\lambda^{d-1}\xi_\lambda) &= \lambda^{-1}i\hat{\xi}_{-d} + (i\hat{\xi}_{1-d})_{\mathfrak{k}} - \lambda i\overline{\hat{\xi}_{-d}}.
\end{cases}
\tag{8.3}
$$

We will use the notation

$$
\begin{aligned}
\alpha_1(\xi_\lambda) &= r(\lambda^{d-1}\xi_\lambda), & \beta_1(\xi_\lambda) &= j(\lambda^{d-1}\xi_\lambda), \\
\alpha_2(\xi_\lambda) &= r(i\lambda^{d-1}\xi_\lambda), & \beta_2(\xi_\lambda) &= j(i\lambda^{d-1}\xi_\lambda).
\end{aligned}
$$

Lemma 8.1

  i) *The vector fields X_1, X_2 are tangent to $\Lambda^d \mathfrak{g}_\tau$.*

  ii) *Moreover, their flows preserve the quantity*

$$\langle \xi_\lambda, \xi_\lambda \rangle = \frac{1}{2\pi} \int_{S^1} \mathrm{tr}(\xi_\lambda.\xi_\lambda)\, dl(\lambda).$$

So, for $\mathfrak{G} = \mathrm{SO}(n+1)$ or $\mathrm{SU}(n)$, since $\langle \cdot, \cdot \rangle$ is negative definite, the flows of X_1 and X_2 stay in spheres and thus exist for all times.

Proof. i) We first have to check that X_1 and X_2 take values in $\Lambda\mathfrak{g}_\tau$. But note that since $d-1$ is even, r is well defined and we have $r(\lambda^{d-1}\xi_\lambda) \in \Lambda\mathfrak{g}_\tau$ for $\xi_\lambda \in \Lambda\mathfrak{g}_\tau$. Thus the same holds for $X_1(\xi_\lambda)$, $X_2(\xi_\lambda)$ by the Lie algebra property of $\Lambda\mathfrak{g}_\tau$. To show that X_1 and X_2 actually take values in $\Lambda^d\mathfrak{g}_\tau$, we use the observation (8.3). Taking the Lie bracket of one of the expressions on the right hand side of (8.3) and ξ_λ yields only powers of λ between $-d$ and d. Another way to prove this is to take into account that

$$\beta_1(\xi_\lambda), \beta_2(\xi_\lambda) \in \Lambda^+_{\mathfrak{b}_\mathfrak{R}}\mathfrak{g}_\tau^{\mathbb{C}}$$

for $\xi_\lambda \in \Lambda^d\mathfrak{g}_\tau$. Hence the Lie bracket of ξ_λ and $\beta_1(\xi_\lambda)$ or $\beta_2(\xi_\lambda)$ contains only powers of λ no less than $-d$. Since $X_1(\xi_\lambda)$ and $X_2(\xi_\lambda)$ are real, we have even $X_1(\xi_\lambda), X_2(\xi_\lambda) \in \Lambda^d\mathfrak{g}_\tau$.

ii) This property is easy to check by a direct computation. $\qquad\square$

Lemma 8.2 *The vector fields X_1 and X_2 commute, i. e. $[X_1, X_2] = 0$.*

Proof. Since α_1 and α_2 are linear, we have

$$
\begin{aligned}
[X_1, X_2](\xi_\lambda) &= [X_1 \cdot \xi_\lambda, \alpha_2(\xi_\lambda)] + [\xi_\lambda, X_1 \cdot \alpha_2(\xi_\lambda)] \\
&\quad - [X_2 \cdot \xi_\lambda, \alpha_1(\xi_\lambda)] - [\xi_\lambda, X_2 \cdot \alpha_1(\xi_\lambda)] \\
&= [[\xi_\lambda, \alpha_1(\xi_\lambda)], \alpha_2(\xi_\lambda)] + [\xi_\lambda, \alpha_2([\xi_\lambda, \alpha_1(\xi_\lambda)])] \\
&\quad - [[\xi_\lambda, \alpha_2(\xi_\lambda)], \alpha_1(\xi_\lambda)] - [\xi_\lambda, \alpha_1([\xi_\lambda, \alpha_2(\xi_\lambda)])].
\end{aligned}
$$

Now use the Jacobi identity for Lie algebras to conclude

$$[X_1, X_2](\xi_\lambda) = [\xi_\lambda, K_\lambda],$$

where

$$K_\lambda = \alpha_2([\xi_\lambda, \alpha_1(\xi_\lambda)]) - \alpha_1([\xi_\lambda, \alpha_2(\xi_\lambda)]) + [\alpha_1(\xi_\lambda), \alpha_2(\xi_\lambda)].$$

We substitute $\alpha_1(\xi_\lambda) = \lambda^{d-1}\xi_\lambda - \beta_1(\xi_\lambda)$ and $\alpha_2(\xi_\lambda) = i\lambda^{d-1}\xi_\lambda - \beta_2(\xi_\lambda)$ and obtain

$$
\begin{aligned}
K_\lambda &= -\alpha_2([\xi_\lambda, \beta_1(\xi_\lambda)] + \alpha_1([\xi_\lambda, \beta_2(\xi_\lambda)]) + [\lambda^{d-1}\xi_\lambda - \beta_1(\xi_\lambda), i\lambda^{d-1}\xi_\lambda - \beta_2(\xi_\lambda)] \\
&= -i\lambda^{d-1}[\xi_\lambda, \beta_1(\xi_\lambda)] + \beta_2([\xi_\lambda, \beta_1(\xi_\lambda)]) + \lambda^{d-1}[\xi_\lambda, \beta_2(\xi_\lambda)] \\
&\quad - \beta_1([\xi_\lambda, \beta_2(\xi_\lambda)]) - \lambda^{d-1}[\xi_\lambda, \beta_2(\xi_\lambda)] + i\lambda^{d-1}[\xi_\lambda, \beta_1(\xi_\lambda)] \\
&\quad + [\beta_1(\xi_\lambda), \beta_2(\xi_\lambda)] \\
&= \beta_2([\xi_\lambda, \beta_1(\xi_\lambda)]) - \beta_1([\xi_\lambda, \beta_2(\xi_\lambda)]) + [\beta_1(\xi_\lambda), \beta_2(\xi_\lambda)].
\end{aligned}
$$

Here we use the fact that $\Lambda\mathfrak{g}_\tau$ and $\Lambda^+_{\mathfrak{b}_\mathfrak{R}}\mathfrak{g}^{\mathbb{C}}_\tau$ are Lie algebras: since the ranges of α_1, α_2 and of β_1, β_2 are $\Lambda\mathfrak{g}_\tau$ and $\Lambda^+_{\mathfrak{b}_\mathfrak{R}}\mathfrak{g}^{\mathbb{C}}_\tau$, respectively, we see from these two representations of K_λ, that

$$K_\lambda \in \Lambda\mathfrak{g}_\tau \cap \Lambda^+_{\mathfrak{b}_\mathfrak{R}}\mathfrak{g}^{\mathbb{C}}_\tau = \{0\}.$$

Thus $[X_1, X_2] = 0$. $\square$

Now, because of Lemma 8.2, we can simultaneously integrate X_1 and X_2 to construct a map

$$\eta_\lambda \colon \mathbb{R}^2 \to \Lambda^d\mathfrak{g}_\tau,$$

such that

$$\begin{cases} \dfrac{\partial\eta_\lambda}{\partial x} &= X_1(\eta_\lambda), \\[2mm] \dfrac{\partial\eta_\lambda}{\partial y} &= X_2(\eta_\lambda). \end{cases}$$

We denote $\alpha_\lambda := r(\lambda^{d-1}\eta_\lambda\,dz)$ and $\beta_\lambda := j(\lambda^{d-1}\eta_\lambda\,dz)$. Then we have the *Lax equation*

$$d\eta_\lambda = [\eta_\lambda, r(\lambda^{d-1}\eta_\lambda\,dz)] = [\eta_\lambda, \alpha_\lambda], \tag{8.4}$$

as can easily be seen from the definition of η_λ. Moreover,

$$\alpha_\lambda = \lambda^{-1}\hat{\eta}_{-d}\,dz + (\hat{\eta}_{1-d}\,dz)_{\mathfrak{k}} + \lambda\overline{\hat{\eta}_{-d}}\,d\bar{z},$$

so we have a representation of α_λ according to our usual decomposition. But we have also the following.

Lemma 8.3 *The 1-forms α_λ and β_λ satisfy*

$$d\alpha_\lambda + \frac{1}{2}[\alpha_\lambda \wedge \alpha_\lambda] = 0,$$

$$d\beta_\lambda - \frac{1}{2}[\beta_\lambda \wedge \beta_\lambda] = 0.$$

Hence Lemma 5.2 and Proposition 7.1 may be applied to prove the existence of an $F_\lambda \colon \Omega \to \mathfrak{G}$, such that $dF_\lambda = F_\lambda\alpha_\lambda$, and such that $u_\lambda = [F_\lambda]$ is harmonic.

Proof. Recall $\beta_\lambda = j(\lambda^{d-1}\eta_\lambda\,dz)$. We know that

$$\begin{aligned} d\eta_\lambda + [\alpha_\lambda, \eta_\lambda] &= 0, \\ d\eta_\lambda - [\beta_\lambda, \eta_\lambda] &= 0. \end{aligned}$$

We may write $\beta_\lambda = \lambda^{d-1}\eta_\lambda\,dz - \alpha_\lambda$. Hence

$$d\alpha_\lambda + \frac{1}{2}[\alpha_\lambda \wedge \alpha_\lambda] + d\beta_\lambda - \frac{1}{2}[\beta_\lambda \wedge \beta_\lambda] = -\lambda^{d-1}(d\eta_\lambda + [\alpha_\lambda, \eta_\lambda])\left(\frac{\partial}{\partial\bar{z}}\right)dz \wedge d\bar{z} = 0.$$

Since we have

$$d\alpha_\lambda + \frac{1}{2}[\alpha_\lambda \wedge \alpha_\lambda] \in \Lambda\mathfrak{g}_\tau$$

and

$$d\beta_\lambda - \frac{1}{2}[\beta_\lambda \wedge \beta_\lambda] \in \Lambda^+_{\mathfrak{b}_\mathfrak{K}}\mathfrak{g}^\mathbb{C}_\tau,$$

this means that both terms vanish because $\Lambda\mathfrak{g}_\tau \cap \Lambda^+_{\mathfrak{b}_K}\mathfrak{g}^\mathbb{C}_\tau = \{0\}$. □

We have thus indeed constructed harmonic maps by integrating a pair of vector fields on the finite dimensional vector spaces $\Lambda^d\mathfrak{g}_\tau$. Such maps are named *finite type* harmonic maps.

8.4 Some further properties of finite type solutions

i) The algorithm developed above does not only create α_λ, but also a 1-form $\beta_\lambda \in T^*\Omega \otimes \Lambda\mathfrak{g}^\mathbb{C}_\tau$, with the property

$$d\beta_\lambda - \frac{1}{2}[\beta_\lambda \wedge \beta_\lambda] = 0.$$

This is—up to the sign—the property requested in lemma 5.2. So in this case the same procedure provides a map $B_\lambda \colon \Omega \to \Lambda\mathfrak{G}^\mathbb{C}_\tau$, such that

$$dB_\lambda = \beta_\lambda B_\lambda,$$

and it is unique up to a prescribed value at some point $z_0 \in \Omega$. We may e. g. choose $B_\lambda(0) = \mathbf{1}_\mathfrak{G}$. Similarly, we may assume that $F_\lambda(0) = \mathbf{1}_\mathfrak{G}$. (Assume $0 \in \Omega$.)

In fact, we have even $\beta_\lambda \in T^*\Omega \otimes \Lambda^+_{\mathfrak{b}_\mathfrak{K}}\mathfrak{g}^\mathbb{C}_\tau$, that is,

$$\beta_\lambda = \sum_{k=0}^\infty \hat{\beta}_k \lambda^k, \quad \hat{\beta}_0 \in \mathfrak{b}_\mathfrak{K}.$$

Hence, if we define the loop group

$$\Lambda^+_{\mathfrak{B}_\mathfrak{K}}\mathfrak{G}^\mathbb{C}_\tau = \{[\lambda \mapsto g_\lambda] \in \Lambda\mathfrak{G}^\mathbb{C}_\tau \colon g_\lambda = \sum_{k \geq 0} \hat{g}_k \lambda^k, \ \hat{g}_0 \in \mathfrak{B}_\mathfrak{K}\},$$

where $\mathfrak{K}^\mathbb{C} = \mathfrak{K}.\mathfrak{B}_\mathfrak{K}$, we can check easily that the Lie algebra of $\Lambda^+_{\mathfrak{B}_\mathfrak{K}}\mathfrak{G}^\mathbb{C}_\tau$ is $\Lambda^+_{\mathfrak{b}_\mathfrak{k}}\mathfrak{g}^\mathbb{C}_\tau$. Thus we obtain

$$B_\lambda \in \Lambda^+_{\mathfrak{B}_\mathfrak{K}}\mathfrak{G}^\mathbb{C}_\tau.$$

ii) Consider once more the equation

$$d\eta_\lambda + [\alpha_\lambda, \eta_\lambda] = 0,$$

where $\alpha_\lambda = F_\lambda^{-1} dF_\lambda$. It implies that

$$F_\lambda^{-1} d(F_\lambda \eta_\lambda F_\lambda^{-1}) F_\lambda = d\eta_\lambda + [\alpha_\lambda, \eta_\lambda] = 0.$$

So $\overset{\circ}{\eta}_\lambda := F_\lambda \eta_\lambda F_\lambda^{-1}$ is a constant in $\Lambda \mathfrak{g}_\tau$.

iii) Furthermore, we have

$$\lambda^{d-1} \eta_\lambda \, dz = \alpha_\lambda + \beta_\lambda,$$

a fact that can also be expressed in the form

$$\begin{aligned}
\lambda^{d-1} \overset{\circ}{\eta}_\lambda \, dz &= \lambda^{d-1} F_\lambda \eta_\lambda F_\lambda^{-1} \, dz = F_\lambda (F_\lambda^{-1} dF_\lambda + dB_\lambda B_\lambda^{-1}) F_\lambda^{-1} \\
&= d(F_\lambda B_\lambda) . (F_\lambda B_\lambda)^{-1}.
\end{aligned}$$

But this means that

$$F_\lambda B_\lambda = e^{\lambda^{d-1} \overset{\circ}{\eta}_\lambda z}. \tag{8.5}$$

iv) Write $\overset{\circ}{\eta}_\lambda$ in the form

$$\overset{\circ}{\eta}_\lambda = e^{-\lambda^{d-1} \overset{\circ}{\eta}_\lambda z} \overset{\circ}{\eta}_\lambda \, e^{\lambda^{d-1} \overset{\circ}{\eta}_\lambda z} = B_\lambda^{-1} F_\lambda^{-1} . \overset{\circ}{\eta}_\lambda \, F_\lambda B_\lambda = B_\lambda^{-1} \eta_\lambda B_\lambda.$$

We know that $\eta_\lambda \in \Lambda^d \mathfrak{g}_\tau$, i. e.

$$\eta_\lambda = \sum_{k=-d}^{d} \hat{\eta}_k \lambda^k.$$

On the other hand, since $B_\lambda \in \Lambda^+_{\mathfrak{B}_{\mathfrak{R}}} \mathfrak{G}^{\mathbb{C}}_\tau$, it has the representation

$$B_\lambda = \sum_{k \geq 0} \hat{B}_k \lambda^k.$$

Hence in the expression $B_\lambda^{-1} \eta_\lambda B_\lambda$, only powers λ^k of λ with $k \geq -d$ appear. But the reality condition $\overset{\circ}{\eta}_\lambda \in \Lambda \mathfrak{g}_\tau$ implies then that in fact $\overset{\circ}{\eta}_\lambda \in \Lambda^d \mathfrak{g}_\tau$.

9 Constant mean curvature tori are of finite type

Recall that CMC surfaces in $\mathbb{R}^3$ correspond to harmonic maps into S^2 by the Gauss map, which are neither holomorphic, nor antiholomorphic. We shall see now that for CMC tori, all such harmonic maps are of finite type, a result of U. Pinkall and I. Sterling. This result can be generalized to harmonic maps from torus into Lie groups [21] or more generally into symmetric spaces [21], [22].

9.1 The result

Theorem 9.1 *[68] All CMC immersions $T^2 \to \mathbb{R}^3$ are of finite type.*

Sketch of the proof (details are in the Appendix)

i) Any immersion of the torus T^2 in $\mathbb{R}^3$ defines a unique conformal structure on T^2 and so can be seen as a conformal immersion of $\mathbb{R}^2/a\mathbb{Z} + b\mathbb{Z}$, where (a, b) is some basis of $\mathbb{R}^2$. Such an immersion can be lifted into a conformal one $X : \mathbb{R}^2 \longrightarrow \mathbb{R}^3$, periodic with period vectors a and b.

Now the Hopf differential of the immersion has the form $f(z)(dz)^2$, where f is holomorphic and periodic, so is constant by Liouville theorem. Would f be equal to zero, then X would be a totally umbilic immersion, so its image would be a subset of a round sphere S^2_X in $\mathbb{R}^3$ (see the proof of Hopf's Theorem in Chapter 4). But this is not possible [3]. So we are left with the situation where f is a constant different from 0.

After suitable normalizations we see that all we need to do is to work with conformal CMC immersions $X : \mathbb{R}^2 \to \mathbb{R}^3$, where X is periodic (with respect to a lattice $a\mathbb{Z} + b\mathbb{Z}$), such that $H = \frac{1}{2}$, and

$$f = \left|\frac{\partial u}{\partial x}\right|^2 - \left|\frac{\partial u}{\partial y}\right|^2 - 2i\left\langle \frac{\partial u}{\partial x}, \frac{\partial u}{\partial y} \right\rangle = -1$$

for the Gauss map $u : \mathbb{R}^2 \to S^2$.

ii) For any $\lambda \in S^1 \subset \mathbb{C}^\star$, construct a "normalized framing"

$$F_\lambda = \begin{pmatrix} e_{1,\lambda} & e_{2,\lambda} & u_\lambda & X_\lambda \\ 0 & 0 & 0 & 1 \end{pmatrix} : \mathbb{R}^2 \longrightarrow SO(3) \ltimes \mathbb{R}^3$$

---

[3]assume that it would be the case: endow $\mathbb{R}^2/a\mathbb{Z} + b\mathbb{Z}$ with g, the pull-back of the metric on S^2_X by X. Then on the one hand the Gauss curvature K_g of g is positive everywhere, but on the other hand the Gauss-Bonnet formula implies $\int_{T^2} K_g \mathrm{dvol}_g = 0$, a contradiction

of X, such that

$$
A_\lambda = F_\lambda^{-1} dF_\lambda = \frac{\lambda^{-1}}{2}
\begin{pmatrix}
0 & 0 & -\sinh\omega & e^\omega \\
0 & 0 & i\cosh\omega & -ie^\omega \\
\sinh\omega & -i\cosh\omega & 0 & 0 \\
0 & 0 & 0 & 0
\end{pmatrix} dz
$$

$$
+ *d\omega
\begin{pmatrix}
0 & -1 & 0 & 0 \\
1 & 0 & 0 & 0 \\
0 & 0 & 0 & 0 \\
0 & 0 & 0 & 0
\end{pmatrix}
\tag{9.1}
$$

$$
+ \frac{\lambda}{2}
\begin{pmatrix}
0 & 0 & -\sinh\omega & e^\omega \\
0 & 0 & -i\cosh\omega & ie^\omega \\
\sinh\omega & i\cosh\omega & 0 & 0 \\
0 & 0 & 0 & 0
\end{pmatrix} d\bar{z}
$$

$$
=
\begin{pmatrix}
0 & -*d\omega & -\sinh\omega\, dx_\lambda & e^\omega\, dx_\lambda \\
*d\omega & 0 & -\cosh\omega\, dy_\lambda & e^\omega\, dy_\lambda \\
\sinh\omega\, dx_\lambda & \cosh\omega\, dy_\lambda & 0 & 0 \\
0 & 0 & 0 & 0
\end{pmatrix},
$$

where $x_\lambda = \frac{1}{2}(\lambda^{-1}z + \lambda\bar{z})$ and $y_\lambda = \frac{1}{2i}(\lambda^{-1}z - \lambda\bar{z})$. (Cf. Chapter 5.) We have then the associated family of CMC immersions X_λ with mean curvature $H = \frac{1}{2}$ and with $f = -\lambda$. The function $\omega\colon \mathbb{R}^2 \to \mathbb{R}$ satisfies

$$
\Delta\omega + \cosh\omega \sinh\omega = 0. \tag{9.2}
$$

The map F_λ may be seen as an application with values in the loop group $\Lambda SO(3) \ltimes \mathbb{R}^3$. Moreover let

$$
P :=
\begin{pmatrix}
1 & 0 & 0 & 0 \\
0 & 1 & 0 & 0 \\
0 & 0 & -1 & 0 \\
0 & 0 & 0 & -1
\end{pmatrix},
$$

and

$$
\tau\colon \quad M(4,\mathbb{R}) \longrightarrow M(4,\mathbb{R})
$$
$$
M \longmapsto Ad_P M = PMP^{-1}.
$$

Then τ is an automorphism of $\Lambda SO(3) \ltimes \mathbb{R}^3$. We also remark that

$$
\tau(A_\lambda) = A_{-\lambda},
$$

so that A_λ has his coefficients in

$$
\Lambda so(3) \oplus \mathbb{R}^3_\tau := \{\lambda \longmapsto \xi_\lambda \in \Lambda so(3) \oplus \mathbb{R}^3 / \lambda \in S^1, \tau(\xi_\lambda) = \xi_{-\lambda}\}.
$$

This is obviously a sub-Lie algebra of $\Lambda so(3) \oplus \mathbb{R}^3$, and the Lie algebra of

$$\Lambda SO(3) \ltimes \mathbb{R}^3_\tau := \{\lambda \longmapsto G_\lambda \in \Lambda SO(3) \ltimes \mathbb{R}^3 / \lambda \in S^1, \tau(G_\lambda) = G_{-\lambda}\}.$$

We deduce that if we choose $F_\lambda(0) \in \Lambda SO(3) \ltimes \mathbb{R}^3_\tau$ (for instance the identity), then F_λ takes its values in $\Lambda SO(3) \ltimes \mathbb{R}^3_\tau$. We shall denote $\mathcal{E}$ the set of maps $F_\lambda : \mathbb{R}^2 \longrightarrow \Lambda SO(3) \ltimes \mathbb{R}^3_\tau$ such that (9.1) is satisfied, for some function $\omega : \mathbb{R}^2 \longrightarrow \mathbb{R}$ which is a solution of (9.2) (F_λ isn't required to be periodic here).

iii) Study deformations of the given $F_\lambda \in \mathcal{E}$ that remain within $\mathcal{E}$. Suppose that the surface X_λ is thereby transformed into

$$X_\lambda + \epsilon \psi_\lambda u_\lambda + o(\epsilon). \tag{9.3}$$

What conditions are to be required of ψ_λ such that this is still a CMC surface with $H = \frac{1}{2}$? Of course the surface given by (9.3) is no longer conformaly parametrized. This problem may however be avoided by considering instead the deformation

$$X_\lambda + \epsilon(t^1_\lambda e_{1,\lambda} + t^2_\lambda e_{2,\lambda} + \psi_\lambda u_\lambda) + o(\epsilon)$$

for suitable t^1_λ, t^2_λ. This amounts to deforming F_λ into

$$F_\lambda + \epsilon F_\lambda T_\lambda + o(\epsilon) \in \mathcal{E},$$

where

$$T_\lambda = \begin{pmatrix} 0 & t^1_{2,\lambda} & t^1_{3,\lambda} & t^1_\lambda \\ t^2_{1,\lambda} & 0 & t^2_{3,\lambda} & t^2_\lambda \\ t^3_{1,\lambda} & t^3_{2,\lambda} & 0 & t^3_\lambda \\ 0 & 0 & 0 & 0 \end{pmatrix}$$

and $t^3_\lambda = \psi_\lambda$. Note that T_λ is completely determined by ψ_λ up to some constant (see in the Appendix). One of the compatibility conditions on ψ_λ that we are looking for turns out to be

$$\Delta\psi_\lambda + \psi_\lambda \cosh 2\omega = 0, \tag{9.4}$$

miraculously (!) the linearization of (9.3) (see Theorem 9.2 in the Appendix).

iv) By this deformation, ω is changed into

$$\omega + \epsilon\dot{\omega} + o(\epsilon).$$

One can compute that

$$\dot{\omega} = \frac{1}{4}\left(\chi_\lambda + \overline{\chi_\lambda} - 2\psi_\lambda\right),$$

where χ_λ is some function constructed starting from ψ_λ by the following algorithm: we let ϕ_λ be a solution of

$$\begin{cases} \dfrac{\partial \phi_\lambda}{\partial z} &= 4\lambda \omega_z \psi_\lambda, \\[2mm] \dfrac{\partial \phi_\lambda}{\partial \bar{z}} &= -\lambda \dfrac{\sinh 2\omega}{2} \psi_\lambda, \end{cases} \tag{9.5}$$

and we set

$$\chi_\lambda := 4\left(\lambda^2 \frac{\partial^2 \psi_\lambda}{(\partial z)^2} - \lambda \omega_z \phi_\lambda\right). \tag{9.6}$$

The existence of solutions to system (9.5) is guaranteed because

$$4(\psi_\lambda \omega_z)_{\bar{z}} + \left(\psi_\lambda \frac{\sinh 2\omega}{2}\right)_z = 0,$$

consequence of (9.2) and (9.4). If χ_λ is constructed from ψ_λ by means of (9.5) and (9.6) we shall write $\psi_\lambda \rightleftharpoons \chi_\lambda$. An important property is that if $\psi_\lambda \rightleftharpoons \chi_\lambda$, then χ_λ is also a solution of (9.4). Note that the solution χ_λ of (9.5) and (9.6) is unique up to some constant (which may depend on λ) times ω_z.

v) There is now an easy way to construct an infinite sequence of formal complex solutions of (9.4). Namely: we start from $\psi_\lambda^{(0)} = 0$ and then we build recursively a sequence $\left(\psi_\lambda^{(n)}\right)_{n \in \mathbb{N}}$ of solutions of (9.4) such that $\psi_\lambda^{(n)} \rightleftharpoons \psi_\lambda^{(n+1)}$, $\forall n \in \mathbb{N}$. It is possible to choose these maps in the form $\psi_\lambda^{(n)} = \lambda^{2n-2} P^{(n)}[\omega]$, where each $P^{(n)}[\omega]$ is a polynomial in ω_z, ω_{zz}, ω_{zzz}, etc ... homogeneous in $\frac{\partial}{\partial z}$ of degree $2n-1$. These functions can in principle be explicitly computed, the sequence begins by

$$\begin{aligned} \psi_\lambda^{(0)} &= \lambda^{-2} P^{(0)}[\omega] = 0, \\ \psi_\lambda^{(2)} &= \lambda^0 P^{(1)}[\omega] = \omega_z, \\ \psi_\lambda^{(3)} &= \lambda^2 P^{(2)}[\omega] = 4\lambda^2(\omega_{zzz} - 2\omega_z^3), \\ \psi_\lambda^{(4)} &= \lambda^4 P^{(3)}[\omega] = 4^2 \lambda^4(\omega_{zzzzz} - 10\omega_{zzz}\omega_z^2 - 10\omega_{zz}^2\omega_z + 6\omega_z^5), \end{aligned}$$

etc.

vi) After some algebraic manipulations, we construct an infinite family of infinitesimal deformations $T_\lambda^{(n,b)} \in T_{F_\lambda}\mathcal{E}$, where $n \in \mathbb{N}$, $b \in \mathbb{C}$, characterized by

$$t_\lambda^{(n,b),3} = \Psi_\lambda^{(n,b)} = b\sum_{k=1}^{n} k\lambda^{-2k} P^{(n-k+1)}[\omega] + \bar{b}\sum_{k=1}^{n} k\lambda^{2k} \overline{P^{(n-k+1)}[\omega]},$$

and the deformation rate of ω induced by $T_\lambda^{(n,b)}$ is

$$\dot{\omega}_\lambda^{(n,b)} = \frac{1}{4}\left(bP^{(n+1)}[\omega] + \overline{bP^{(n+1)}[\omega]}\right).\tag{9.7}$$

In particular this expression does not depend on λ.

vii) This is the only analysis part in this process. Assume everything is periodic, i. e. we are working on a torus. Then the linear equation (9.4), which is an eigenvalue problem for the Laplace operator, has a finite dimensional space of solutions. In particular, there exists an integer N, such that

$$\mathrm{rank}_{\mathbb{C}}\{P^{(1)}[\omega],\dots,P^{(N)}[\omega]\} = \mathrm{rank}_{\mathbb{C}}\{P^{(1)}[\omega],\dots,P^{(N+1)}[\omega]\} = N.$$

viii) We now choose $(b_1,\dots,b_{N+1}) \in \mathbb{C}^{N+1} \setminus \{0\}$ (in particular $b_{N+1} \neq 0$) such that

$$b_1 P^{(1)}[\omega] + \cdots + b_{N+1}P^{(N+1)}[\omega] = 0 \ \text{ on } \mathbb{R}^2.\tag{9.8}$$

Then we let

$$T_\lambda^\star := T_\lambda^{(0,b_1)} + T_\lambda^{(1,b_2)} + \cdots + T_\lambda^{(N,b_{N+1})}.$$

It is an infinitesimal deformation in $T_{F_\lambda}\mathcal{E}$ which does not vanish. Indeed the family $\left(\lambda^{-2}P^{(1)}[\omega],\dots,\sum_{k=1}^N k\lambda^{-2k}P^{(N-k+1)}[\omega]\right)$ is linearly independant and so $t_\lambda^{\star,3} = \Psi_\lambda^\star := \Psi_\lambda^{(1,b_2)} + \cdots + \Psi_\lambda^{(N,b_{N+1})}$ does not vanish.

ix) *Think and interpret* : Let

$$F_\lambda + \epsilon F_\lambda T_\lambda^\star + o(\epsilon) \in \mathcal{E}$$

be the infinitesimal deformation of F_λ induced given by $T_\lambda^\star$. The deformation rate of ω induced by $T_\lambda^\star$ is $\dot{\omega}_\lambda^{(0,b_1)} + \cdots + \dot{\omega}_\lambda^{(N,b_{N+1})}$ (see (9.7)). But because of (9.8) this is just zero: the surface is not deformed at all, but just moved around in $\mathbb{R}^3$. So

$$F_\lambda T_\lambda^\star = D_\lambda F_\lambda\tag{9.9}$$

for some constant D_λ in the Lie algebra of $\Lambda SO(3) \ltimes \mathbb{R}^3$.

x) We conclude that $T_\lambda^\star = F_\lambda^{-1}D_\lambda F_\lambda$ satisfies

$$dT_\lambda^\star + [A_\lambda, T_\lambda^\star] = 0.$$

If we write

$$T_\lambda^\star = \begin{pmatrix} \check{T}_\lambda^\star & t_\lambda^\star \\ 0 & 0 \end{pmatrix}, \quad A_\lambda = \begin{pmatrix} \check{A}_\lambda & a_\lambda \\ 0 & 0 \end{pmatrix},$$

then we have

$$d\check{T}_\lambda^\star + [\check{A}_\lambda, \check{T}_\lambda^\star] = 0.$$

But by construction, this includes only a finite number of powers of λ (i. e. $\check{T}_\lambda^\star, \check{A}_\lambda$ are polynomial in λ). Hence we have a finite type representation of X, as claimed.

9.2 Appendix

In this Appendix we expose the technical details of the proof of Theorem 8.1. We shall use the notations

$$\Lambda\mathbb{R}_0 := \{\lambda \longmapsto t_\lambda \in \mathbb{R}/t_{-\lambda} = t_\lambda\},$$

$$\Lambda\mathbb{R}_1 := \{\lambda \longmapsto t_\lambda \in \mathbb{R}/t_{-\lambda} = -t_\lambda\}.$$

Study of the deformation of F_λ

Given some extended frame $F_\lambda \in \mathcal{E}$ we want to study infinitesimal deformations $F_{\epsilon,\lambda} := F_\lambda + \epsilon F_\lambda T_\lambda + o(\epsilon)$ where T_λ is a smooth map from $\mathbb{R}^2$ to $\Lambda SO(3) \ltimes \mathbb{R}^3_\tau$ and such that $F_{\epsilon,\lambda}$ still belongs to $\mathcal{E}$. Thus T_λ may be viewed as an element of $T_{F_\lambda}\mathcal{E}$, the set of tangent vectors to $\mathcal{E}$ at F_λ. We shall denote

$$T_\lambda = \begin{pmatrix} 0 & t^1_{2,\lambda} & t^1_{3,\lambda} & t^1_\lambda \\ t^2_{1,\lambda} & 0 & t^2_{3,\lambda} & t^2_\lambda \\ t^3_{1,\lambda} & t^3_{2,\lambda} & 0 & t^3_\lambda \\ 0 & 0 & 0 & 0 \end{pmatrix}$$

and sometimes drop the λ when there is no ambiguity. The question is to characterize all such T_λ's. In view of (9.1) the answer is simple: one considers any one-parameter deformation of ω, $(\omega_\epsilon)_{\epsilon\in(-\epsilon_0,\epsilon_0)}$ with $\omega_0 = \omega$ such that for any ϵ, ω_ϵ is a solution of

$$\Delta\omega_\epsilon + \sinh\omega_\epsilon \cosh\omega_\epsilon = 0.$$

Then

$$A_{\epsilon,\lambda} := \begin{pmatrix} 0 & -\star d\omega_\epsilon & -\sinh\omega_\epsilon dx_\lambda & e^{\omega_\epsilon}dx_\lambda \\ \star d\omega_\epsilon & 0 & -\cosh\omega_\epsilon dy_\lambda & e^{\omega_\epsilon}dy_\lambda \\ \sinh\omega_\epsilon dx_\lambda & \cosh\omega_\epsilon dy_\lambda & 0 & 0 \\ 0 & 0 & 0 & 0 \end{pmatrix} \qquad (9.10)$$

is integrable, i.e. $dA_{\epsilon,\lambda} + A_{\epsilon,\lambda} \wedge A_{\epsilon,\lambda} = 0$ and integration of the equation $dF_{\epsilon,\lambda} = F_{\epsilon,\lambda}A_{\epsilon,\lambda}$ leads to the family of deformations. Notice that if one set $\omega_\epsilon := \omega + \epsilon\dot\omega + o(\epsilon)$, then $\dot\omega$ is a solution of the linearized sinh-Gordon equation

$$\Delta\dot\omega + \dot\omega \cosh 2\omega = 0.$$

A crucial point - which is remarked by U. Pinkall and I. Sterling in [68] - is that one may also construct a T_λ in $T_{F_\lambda}\mathcal{E}$ knowing its component t^3_λ (modulo some constant), a map into $\Lambda\mathbb{R}_0$.

Theorem 9.2 *a) Let $T_\lambda \in T_{F_\lambda}\mathcal{E}$, then the component t_λ^3 of T_λ is a solution of the linearized sinh-Gordon equation*

$$\Delta t_\lambda^3 + t_\lambda^3 \cosh 2\omega = 0. \tag{9.11}$$

b) Let $\psi_\lambda : \mathbb{R}^2 \longrightarrow \Lambda\mathbb{R}_0$. Then the overdetermined linear system

$$\begin{cases} \dfrac{\partial \phi_\lambda}{\partial z} &= 4\lambda \dfrac{\partial \omega}{\partial z} \dfrac{\partial \psi_\lambda}{\partial z} \\ \dfrac{\partial \phi_\lambda}{\partial \overline{z}} &= -\lambda \dfrac{\sinh 2\omega}{2} \psi_\lambda \end{cases} \tag{9.12}$$

has a solution if and only if ψ_λ is a solution of the linearized sinh-Gordon equation (9.11).

c) (A partial converse to a)) Let $\psi_\lambda : \mathbb{R}^2 \longrightarrow \Lambda\mathbb{R}_0$ be a solution of (9.11). Then there exists some $T_\lambda \in T_{F_\lambda}\mathcal{E}$, such that $t_\lambda^3 = \psi_\lambda$ if and only there exists a solution ϕ_λ of (9.12), such that

$$\Omega(\psi_\lambda, \phi_\lambda) := \left(\lambda^2 \frac{\partial^2 \psi_\lambda}{\partial z^2} - \lambda \frac{\partial \omega}{\partial z} \phi_\lambda \right) + \left(\lambda^{-2} \overline{\frac{\partial^2 \psi_\lambda}{\partial z^2}} - \lambda^{-1} \frac{\partial \omega}{\partial \overline{z}} \overline{\phi_\lambda} \right) - \frac{1}{2}\psi_\lambda \tag{9.13}$$

is a function which does not depend on λ.

The proof of Theorem 9.2 relies on the following Lemma.

Lemma 9.1 *Let $T_\lambda \in T_{F_\lambda}\mathcal{E}$ then*

- *the component t_λ^3 of T_λ satisfies the linearized sinh-Gordon equation (9.11)*
- *there exists a map $\phi_\lambda : \mathbb{R}^2 \longrightarrow \Lambda\mathbb{R}_1$ (where $\Lambda\mathbb{R}_1 := \{\lambda \longmapsto t_\lambda \in \mathbb{R}/t_{-\lambda} = -t_\lambda\}$), which is a solution of (9.12), with $\psi_\lambda = t_\lambda^3$, such that*

$$t_\lambda^1 + i t_\lambda^2 = e^\omega \left(2\lambda \frac{\partial \psi_\lambda}{\partial z} - \phi_\lambda \right), \tag{9.14}$$

$$t_{1,\lambda}^3 + i t_{2,\lambda}^3 = e^\omega \left(\lambda \frac{\partial \psi_\lambda}{\partial z} - \frac{\phi_\lambda}{2} \right) + e^{-\omega} \left(\lambda^{-1} \frac{\partial \psi_\lambda}{\partial \overline{z}} - \frac{\overline{\phi_\lambda}}{2} \right), \tag{9.15}$$

$$t_{2,\lambda}^1 = i \left(\lambda^2 \frac{\partial^2 \psi_\lambda}{\partial z^2} - \lambda \frac{\partial \omega}{\partial z} \phi_\lambda \right) - i \left(\lambda^{-2} \overline{\frac{\partial^2 \psi_\lambda}{\partial z^2}} - \lambda^{-1} \frac{\partial \omega}{\partial \overline{z}} \overline{\phi_\lambda} \right), \tag{9.16}$$

$$\dot{\omega} = \Omega(\psi_\lambda, \phi_\lambda). \tag{9.17}$$

The proof of Lemma 9.1 occupies the end of this section. We shall first prove Theorem 9.2 assuming that Lemma 9.1 is true.

Proof of Theorem 9.2. The property in *a)* is a restatement of part of Lemma 9.1. To prove *b)*, we need to write down the integrability condition of system (9.12), i. e. that

$$d\left(4\lambda\omega_z\frac{\partial\psi_\lambda}{\partial z}dz - \lambda\frac{\sinh 2\omega}{2}\psi_\lambda d\bar{z}\right) = 0.$$

(Here we use the subscript ω_z for $\frac{\partial\omega}{\partial z}$, etc...). Using $\omega_{z\bar{z}} + \frac{1}{8}\sinh 2\omega = 0$, this simplifies into

$$\lambda\omega_z\left(4\frac{\partial^2\psi_\lambda}{\partial z\partial\bar{z}} + \psi_\lambda\cosh 2\omega\right)d\bar{z}\wedge dz = 0,$$

which is clearly equivalent to (9.11).

Lastly for proving *c)* let us consider some ψ_λ, which is a solution of (9.11).

Let us assume first that there exists a $T_\lambda \in T_{F_\lambda}\mathcal{E}$ such that $t_1^3 = \psi_\lambda$. Then the infinitesimal deformation $\dot{\omega}$ of ω induced by T_λ needs to be independent of λ. But because of (9.17), in Lemma 9.1, it means exactly that $\Omega(\psi_\lambda, \phi_\lambda)$ is independent of λ, for a suitable choice of ϕ_λ.

Conversely if there exists some solution ϕ_λ of (9.12), such that $\Omega(\psi_\lambda, \phi_\lambda)$ is independent of λ, then formulas (9.14), (9.15) and (9.16) defines a map T_λ. A long computation (which basically uses the material of the proof of Lemma 9.1) shows that actually $T_\lambda \in T_{F_\lambda}\mathcal{E}$. □

Proof of Lemma 9.1
Step 1: Writing the relations satisfied by T_λ
One defines $F_{\epsilon,\lambda} := F_\lambda + \epsilon F_\lambda T_\lambda + o(\epsilon) \in \mathcal{E}$ and one writes that

$$F_{\epsilon,\lambda}^{-1}dF_{\epsilon,\lambda} = A_{\epsilon,\lambda}, \tag{9.18}$$

where $A_{\epsilon,\lambda}$ has been defined in (9.10). The left hand side of (9.18) is

$$F_{\epsilon,\lambda}^{-1}dF_{\epsilon,\lambda} = A_\lambda + \epsilon(dT_\lambda + [A_\lambda, T_\lambda]) + o(\epsilon),$$

where

$$dT_\lambda + [A_\lambda, T_\lambda] =$$

$$\left(\begin{array}{cc}
\begin{array}{c}
0 \\
dt_1^2 - t_3^2\sinh\omega dx_\lambda + t_3^1\cosh\omega dy_\lambda \\
dt_1^3 - t_2^3\star d\omega + t_1^2\cosh\omega dy_\lambda \\
0
\end{array} &
\begin{array}{c}
dt_2^1 - t_2^3\sinh\omega dx_\lambda + t_1^3\cosh\omega dy_\lambda \\
0 \\
dt_2^3 + t_1^3\star d\omega + t_2^1\sinh\omega dx_\lambda \\
0
\end{array} \\
& \\
\begin{array}{c}
dt_3^1 + t_2^3\star d\omega + t_2^1\cosh\omega dy_\lambda \\
dt_3^2 - t_1^3\star d\omega + t_1^2\sinh\omega dx_\lambda \\
0 \\
0
\end{array} &
\begin{array}{c}
dt^1 - t^2\star d\omega - t^3\sinh\omega dx_\lambda - e^\omega t_2^1 dy_\lambda \\
dt^2 + t^1\star d\omega - t^3\cosh\omega dy_\lambda - e^\omega t_1^2 dx_\lambda \\
dt^3 + t^1\sinh\omega dx_\lambda + t^2\cosh\omega dy_\lambda - e^\omega(t_1^3 dx_\lambda + t_2^3 dy_\lambda) \\
0
\end{array}
\end{array}\right).$$

Whereas the right hand side of (9.18) is

$$A_{\epsilon,\lambda} = A_\lambda + \epsilon \begin{pmatrix} 0 & -\star d\dot\omega & -\dot\omega \cosh\omega dx_\lambda & \dot\omega e^\omega dx_\lambda \\ \star d\dot\omega & 0 & -\dot\omega \sinh\omega dy_\lambda & \dot\omega e^\omega dy_\lambda \\ \dot\omega \cosh\omega dx_\lambda & \dot\omega \sinh\omega dy_\lambda & 0 & 0 \\ 0 & 0 & 0 & 0 \end{pmatrix} + o(\epsilon).$$

Thus we are led to the following system of equations.

$$dt^1 - t^2 \star d\omega - t^3 \sinh\omega dx_\lambda - e^\omega t_2^1 dy_\lambda = \dot\omega e^\omega dx_\lambda, \tag{9.19}$$

$$dt^2 + t^1 \star d\omega - t^3 \cosh\omega dy_\lambda + e^\omega t_2^1 dx_\lambda = \dot\omega e^\omega dy_\lambda, \tag{9.20}$$

$$dt_1^3 - t_2^3 \star d\omega - t_2^1 \cosh\omega dy_\lambda = \dot\omega \cosh\omega dx_\lambda, \tag{9.21}$$

$$dt_2^3 + t_1^3 \star d\omega + t_2^1 \sinh\omega dx_\lambda = \dot\omega \sinh\omega dy_\lambda, \tag{9.22}$$

$$dt_2^1 - t_2^3 \sinh\omega dx_\lambda + t_1^3 \cosh\omega dy_\lambda + \star d\dot\omega = 0, \tag{9.23}$$

$$dt^3 + t^1 \sinh\omega dx_\lambda + t^2 \cosh\omega dy_\lambda - e^\omega(t_1^3 dx_\lambda + t_2^3 dy_\lambda) = 0. \tag{9.24}$$

Step 2: Useful relations
Relation (9.24) allows to express t_1^3 and t_2^3 in terms of t^1, t^2 and t^3:

$$\begin{cases} t_1^3 &= e^{-\omega}\left(\dfrac{\partial t^3}{\partial x_\lambda} + t^1 \sinh\omega \right) \\[2mm] t_2^3 &= e^{-\omega}\left(\dfrac{\partial t^3}{\partial y_\lambda} + t^2 \cosh\omega \right). \end{cases} \tag{9.25}$$

Furthermore, relations (9.19), (9.20) and (9.21) can be replaced by their respective projections along $\frac{\partial}{\partial x_\lambda} := \lambda\frac{\partial}{\partial z} + \lambda^{-1}\frac{\partial}{\partial \bar z}$ and $\frac{\partial}{\partial y_\lambda} := i\lambda\frac{\partial}{\partial z} - i\lambda^{-1}\frac{\partial}{\partial \bar z}$:

$$\frac{\partial t^1}{\partial x_\lambda} + t^2 \frac{\partial \omega}{\partial y_\lambda} - t^3 \sinh\omega = \dot\omega e^\omega, \tag{9.26}$$

$$\frac{\partial t^1}{\partial y_\lambda} - t^2 \frac{\partial \omega}{\partial x_\lambda} = t_2^1 e^\omega, \tag{9.27}$$

$$\frac{\partial t^2}{\partial x_\lambda} - t^1 \frac{\partial \omega}{\partial y_\lambda} = -t_2^1 e^\omega, \tag{9.28}$$

$$\frac{\partial t^2}{\partial y_\lambda} + t^1 \frac{\partial \omega}{\partial x_\lambda} - t^3 \cosh\omega = \dot\omega e^\omega, \tag{9.29}$$

$$\frac{\partial t_1^3}{\partial x_\lambda} + t_2^3 \frac{\partial \omega}{\partial y_\lambda} = \dot\omega \cosh\omega, \tag{9.30}$$

$$\frac{\partial t_1^3}{\partial y_\lambda} - t_2^3 \frac{\partial \omega}{\partial x_\lambda} = t_2^1 \cosh \omega, \tag{9.31}$$

$$\frac{\partial t_2^3}{\partial x_\lambda} - t_1^3 \frac{\partial \omega}{\partial y_\lambda} = -t_2^1 \sinh \omega. \tag{9.32}$$

$$\frac{\partial t_2^3}{\partial y_\lambda} + t_1^3 \frac{\partial \omega}{\partial x_\lambda} = \dot\omega \sinh \omega, \tag{9.33}$$

Lastly, two further relations can be deduced from the four first ones of the previous series: substracting (9.26) - (9.29) gives

$$\frac{\partial(e^{-\omega}t^1)}{\partial x_\lambda} - \frac{\partial(e^{-\omega}t^2)}{\partial y_\lambda} = -t^3 e^{-2\omega}, \tag{9.34}$$

and (9.27) + (9.28) gives

$$\frac{\partial(e^{-\omega}t^1)}{\partial y_\lambda} + \frac{\partial(e^{-\omega}t^2)}{\partial x_\lambda} = 0. \tag{9.35}$$

Step 3: We show that t^3 is a solution of the linearized sinh-Gordon equation

Suming (9.30) + (9.33) gives

$$\dot\omega e^{2\omega} = \frac{\partial(e^\omega t_1^3)}{\partial x_\lambda} + \frac{\partial(e^\omega t_2^3)}{\partial y_\lambda},$$

which, using (9.25) to eliminate t_1^3 and t_2^3, leads to

$$\dot\omega e^{2\omega} = \Delta t^3 + \frac{\partial(t^1 \sinh \omega)}{\partial x_\lambda} + \frac{\partial(t^2 \cosh \omega)}{\partial y_\lambda}.$$

Now (9.26) + (9.29) gives

$$\dot\omega e^{2\omega} = \frac{1}{2}\left(\frac{\partial(e^\omega t^1)}{\partial x_\lambda} + \frac{\partial(e^\omega t^2)}{\partial y_\lambda} - t^3 e^{2\omega}\right).$$

These two relations implies by eliminating $\dot\omega e^{2\omega}$

$$\Delta t^3 + \frac{t^3}{2}e^{2\omega} = \frac{1}{2}\left(\frac{\partial(e^{-\omega}t^1)}{\partial x_\lambda} - \frac{\partial(e^{-\omega}t^2)}{\partial y_\lambda}\right).$$

And by using (9.34) one gets

$$\Delta t^3 + t^3 \cosh \omega = 0. \tag{9.36}$$

Step 4: Equations on t^1 and t^2

- We sum $-\sinh\omega$ (9.26) $+\cosh\omega$ (9.29) to obtain

$$-\sinh\omega\left(\frac{\partial t^1}{\partial x_\lambda}+t^2\frac{\partial\omega}{\partial y_\lambda}-t^3\sinh\omega\right)+\cosh\omega\left(\frac{\partial t^2}{\partial y_\lambda}+t^1\frac{\partial\omega}{\partial x_\lambda}-t^3\cosh\omega\right)=\dot\omega,$$

and substracting e^ω (9.30) - e^ω(9.33) gives

$$e^\omega\left(\frac{\partial t_2^3}{\partial y_\lambda}+t_1^3\frac{\partial\omega}{\partial x_\lambda}\right)-e^\omega\left(\frac{\partial t_1^3}{\partial x_\lambda}+t_2^3\frac{\partial\omega}{\partial y_\lambda}\right)=-\dot\omega.$$

Now we sum both equations and we use (9.25) in order to eliminate t_1^3 and t_2^3. It leads to

$$2\frac{\partial t^2}{\partial y_\lambda}\cosh\omega-2\frac{\partial t^1}{\partial x_\lambda}\sinh\omega-2t^2\frac{\partial\omega}{\partial y_\lambda}\cosh\omega+2t^1\frac{\partial\omega}{\partial x_\lambda}\sinh\omega$$
$$-\frac{\partial^2 t^3}{(\partial x_\lambda)^2}+\frac{\partial^2 t^3}{(\partial y_\lambda)^2}+2\frac{\partial\omega}{\partial x_\lambda}\frac{\partial t^3}{\partial x_\lambda}-2\frac{\partial\omega}{\partial y_\lambda}\frac{\partial t^3}{\partial y_\lambda}=t^3.$$

A computation using (9.34) gives

$$\frac{\partial(e^{-\omega}t^1)}{\partial x_\lambda}+\frac{\partial(e^{-\omega}t^2)}{\partial y_\lambda}=e^{2\omega}\left[\frac{\partial}{\partial x_\lambda}\left(e^{-2\omega}\frac{\partial t^3}{\partial x_\lambda}\right)-\frac{\partial}{\partial y_\lambda}\left(e^{-2\omega}\frac{\partial t^3}{\partial y_\lambda}\right)\right]. \quad (9.37)$$

- Similarly the combination $-\sinh\omega$ (9.27) $-\cosh\omega$ (9.28) gives

$$-\sinh\omega\left(\frac{\partial t^1}{\partial y_\lambda}-t^2\frac{\partial\omega}{\partial x_\lambda}\right)-\cosh\omega\left(\frac{\partial t^2}{\partial x_\lambda}-t^1\frac{\partial\omega}{\partial y_\lambda}\right)=t_2^1,$$

and the sum e^ω (9.31) $+e^\omega$ (9.32) gives

$$-e^\omega\left(\frac{\partial t_1^3}{\partial y_\lambda}-t_2^3\frac{\partial\omega}{\partial x_\lambda}\right)-e^\omega\left(\frac{\partial t_2^3}{\partial x_\lambda}-t_1^3\frac{\partial\omega}{\partial y_\lambda}\right)=-t_2^1.$$

We sum these two relations and use (9.25) to obtain

$$-\frac{\partial t^2}{\partial x_\lambda}\cosh\omega-\frac{\partial t^1}{\partial y_\lambda}\sinh\omega+t^2\frac{\partial\omega}{\partial x_\lambda}\cosh\omega+t^1\frac{\partial\omega}{\partial y_\lambda}\sinh\omega$$
$$-\frac{\partial^2 t^3}{\partial x_\lambda\partial y_\lambda}+\frac{\partial\omega}{\partial y_\lambda}\frac{\partial t^3}{\partial x_\lambda}+\frac{\partial\omega}{\partial x_\lambda}\frac{\partial t^3}{\partial y_\lambda}=0.$$

Then a computation using (9.35) gives

$$\frac{\partial(e^{-\omega}t^1)}{\partial y_\lambda}-\frac{\partial(e^{-\omega}t^2)}{\partial x_\lambda}=e^{2\omega}\left[\frac{\partial}{\partial x_\lambda}\left(e^{-2\omega}\frac{\partial t^3}{\partial y_\lambda}\right)+\frac{\partial}{\partial y_\lambda}\left(e^{-2\omega}\frac{\partial t^3}{\partial x_\lambda}\right)\right]. \quad (9.38)$$

We can conclude this step by the following equations which summarize relations (9.34), (9.35), (9.37) and (9.38).

$$\frac{\partial(e^{-\omega}(t^1 + it^2))}{\partial z_\lambda} = 2\frac{\partial^2 t^3}{(\partial z_\lambda)^2} - 4\frac{\partial \omega}{\partial z_\lambda}\frac{\partial t^3}{\partial z_\lambda}, \tag{9.39}$$

$$\frac{\partial(e^{-\omega}(t^1 + it^2))}{\partial \overline{z_\lambda}} = -\frac{e^{-2\omega}}{2}t^3, \tag{9.40}$$

where $\frac{\partial}{\partial z_\lambda} = \lambda\frac{\partial}{\partial z}$ and $\frac{\partial}{\partial \overline{z_\lambda}} = \lambda^{-1}\frac{\partial}{\partial \overline{z}}$.

- We shall pose

$$\phi_\lambda := 2\frac{\partial t^3}{\partial z_\lambda} - e^{-\omega}(t^1 + it^2) \Longleftrightarrow e^{-\omega}(t^1 + it^2) = 2\frac{\partial t^3}{\partial z_\lambda} - \phi_\lambda. \tag{9.41}$$

Then relations (9.39) and (9.40) allows to express ϕ_λ in terms of t^3 (also using (9.36)):

$$\frac{\partial \phi_\lambda}{\partial z_\lambda} = 2\frac{\partial^2 t^3}{(\partial z_\lambda)^2} - \frac{\partial}{\partial z_\lambda}\left(e^{-\omega}(t^1 + it^2)\right) = 4\frac{\partial \omega}{\partial z_\lambda}\frac{\partial t^3}{\partial z_\lambda}, \tag{9.42}$$

$$\frac{\partial \phi_\lambda}{\partial \overline{z_\lambda}} = 2\frac{\partial^2 t^3}{\partial z_\lambda \partial \overline{z_\lambda}} - \frac{\partial}{\partial \overline{z_\lambda}}\left(e^{-\omega}(t^1 + it^2)\right) = -\frac{\sinh\omega}{2}t^3. \tag{9.43}$$

Step 5: Computation of $\dot{\omega}$

Suming (9.26) + (9.29) we obtain

$$e^{2\omega}(2\dot{\omega} + t^3) = \frac{\partial(e^\omega(t^1 + it^2))}{\partial z_\lambda} + \frac{\partial(e^\omega(t^1 + it^2))}{\partial \overline{z_\lambda}}.$$

And using identities like

$$\frac{\partial(e^\omega(t^1 + it^2))}{\partial z_\lambda} = e^{2\omega}\frac{\partial(e^{-\omega}(t^1 + it^2))}{\partial z_\lambda} + 2e^{2\omega}\frac{\partial \omega}{\partial z_\lambda}(e^{-\omega}(t^1 + it^2)),$$

we get

$$\dot{\omega} = -\frac{t^3}{2} + \left(\frac{\partial \omega}{\partial z_\lambda}(e^{-\omega}(t^1 + it^2)) + \frac{1}{2}\frac{\partial(e^{-\omega}(t^1 + it^2))}{\partial z_\lambda}\right)$$

$$+ \left(\frac{\partial \omega}{\partial \overline{z_\lambda}}(e^{-\omega}(t^1 - it^2)) + \frac{1}{2}\frac{\partial(e^{-\omega}(t^1 - it^2))}{\partial \overline{z_\lambda}}\right).$$

Now we use (9.41) in order to obtain a formula which relates $\dot{\omega}$ in function of t^3 and ϕ_λ. Simplifications using (9.42) and (9.43) occur and give

$$\dot{\omega} = -\frac{t^3}{2} + \left(\frac{\partial^2 t^3}{(\partial z_\lambda)^2} - \frac{\partial \omega}{\partial z_\lambda}\phi_\lambda\right) + \left(\frac{\partial^2 t^3}{(\partial \overline{z_\lambda})^2} - \frac{\partial \omega}{\partial \overline{z_\lambda}}\overline{\phi_\lambda}\right).$$

We recognize the relation $\dot{\omega} = \Omega(t_\lambda^3, \phi_\lambda)$ where Ω has been defined in Theorem 9.2 by formula (9.13).

Step 6: Conclusion

The map ϕ_λ which was introduced through the relation (9.41) is actually completely determined from $\psi_\lambda := t_\lambda^3$ by relations (9.42) and (9.43), up to a constant. Then the relation (9.41) implies easily relation (9.14). And using (9.25) one deduces (9.15). Lastly one gets from (9.27) and (9.28) that

$$t_2^1 = \frac{e^{-2\omega}}{2}\left(\frac{\partial(e^\omega t^1)}{\partial y_\lambda} - \frac{\partial(e^\omega t^2)}{\partial x_\lambda}\right) = -e^{-2\omega}\mathrm{Im}\frac{\partial(e^\omega(t^1 + it^2))}{\partial z_\lambda},$$

from which we deduce (9.16). $\qquad\square$

A Bäcklund transformation of solutions of the linearized sinh-Gordon equation

In the previous subsection we have seen that for any $\psi_\lambda : \mathbb{R}^2 \longrightarrow \Lambda\mathbb{R}_0$ which is a solution of

$$\Delta\psi_\lambda + \psi_\lambda \cosh 2\omega = 0, \tag{9.44}$$

there exists a solution $\phi_\lambda : \mathbb{R}^2 \longrightarrow \Lambda\mathbb{R}_1$ of

$$\begin{cases} \dfrac{\partial\phi_\lambda}{\partial z} = 4\lambda\dfrac{\partial\omega}{\partial z}\dfrac{\partial\psi_\lambda}{\partial z} \\ \dfrac{\partial\phi_\lambda}{\partial\bar{z}} = -\lambda\dfrac{\sinh 2\omega}{2}\psi_\lambda \end{cases}. \tag{9.45}$$

Obviously ϕ_λ is unique up to a constant in $\Lambda\mathbb{R}_1$.

It will be useful to extend our setting to maps $\psi_\lambda : \mathbb{R}^2 \longrightarrow \Lambda\mathbb{C}_0 := \{\lambda \longmapsto t_\lambda \in \mathbb{C}/t_{-\lambda} = t_\lambda\}$ which are solutions of (9.44). Then there exist solutions $\phi_\lambda : \mathbb{R}^2 \longrightarrow \Lambda\mathbb{C}_1 := \{\lambda \longmapsto t_\lambda \in \mathbb{C}/t_{-\lambda} = -t_\lambda\}$ of (9.45).

Definition 9.1 *Let* $\psi_\lambda : \mathbb{R}^2 \longrightarrow \Lambda\mathbb{C}_0$ *be a solution of (9.44) and* $\chi_\lambda : \mathbb{R}^2 \longrightarrow \Lambda\mathbb{C}_0$. *We write* $\psi_\lambda \rightleftharpoons \chi_\lambda$ *if and only if there exists a solution* $\phi_\lambda : \mathbb{R}^2 \longrightarrow \Lambda\mathbb{C}_1$ *of (9.45) such that*

$$\chi_\lambda = 4\left(\lambda^2\frac{\partial^2\psi_\lambda}{(\partial z)^2} - \lambda\frac{\partial\omega}{\partial z}\phi_\lambda\right).$$

If $\psi_\lambda \rightleftharpoons \chi_\lambda^0$, *the set of maps* χ_λ *such that* $\psi_\lambda \rightleftharpoons \chi_\lambda$ *is exactly* $\{\chi_\lambda^0 + C_\lambda\omega_z/C_\lambda \in \Lambda\mathbb{C}_0\}$.

This definition is motivated by Lemma 9.1: indeed relation (9.17) can be restated as

$$\dot{\omega} = \frac{1}{4}\chi_\lambda + \frac{1}{4}\overline{\chi_\lambda} - \frac{1}{2}\psi_\lambda,$$

for $\psi_\lambda = t^3$ and $\psi_\lambda \rightleftharpoons \chi_\lambda$ [4]. Proposition 9.1 below leads to another charaterization of χ_λ.

---

[4]similarly relation (9.16) can be written as $t_{2,\lambda}^1 = \frac{i}{4}\chi_\lambda - \frac{i}{4}\overline{\chi_\lambda}$

Lemma 9.2 If $\psi_\lambda : \mathbb{R}^2 \longrightarrow \Lambda\mathbb{C}_0$ is a solution of (9.44) and $\psi_\lambda \rightleftharpoons \chi_\lambda$, then χ_λ is also a solution of (9.44).

Proof. It can be checked through a direct computation. Alternatively it follows from Lemma 9.1, since $\dot{\omega}$ and t_λ^3 are both solutions of (9.44) and because of (9.17). Lastly it is also a consequence of the Proposition below.   □

Proposition 9.1 Let $\psi_\lambda : \mathbb{R}^2 \longrightarrow \Lambda\mathbb{C}_0$ is a solution of (9.44). Then the set of maps $\chi_\lambda : \mathbb{R}^2 \longrightarrow \Lambda\mathbb{C}_0$ for which there exists two maps $a_\lambda, b_\lambda : \mathbb{R}^2 \longrightarrow \Lambda\mathbb{C}$ such that

$$
\begin{cases}
\dfrac{\partial}{\partial \bar{z}} \begin{pmatrix} \cosh\omega & -i\sinh\omega \\ i\sinh\omega & \cosh\omega \end{pmatrix} \begin{pmatrix} a_\lambda \\ b_\lambda \end{pmatrix} = \lambda\psi_\lambda \begin{pmatrix} -i\sinh 2\omega \\ \cosh 2\omega \end{pmatrix} \\[4mm]
\dfrac{\partial}{\partial z} \begin{pmatrix} \cosh\omega & i\sinh\omega \\ -i\sinh\omega & \cosh\omega \end{pmatrix} \begin{pmatrix} a_\lambda \\ b_\lambda \end{pmatrix} = -\lambda^{-1}\chi_\lambda \begin{pmatrix} i\sinh 2\omega \\ \cosh 2\omega \end{pmatrix}
\end{cases}
\tag{9.46}
$$

coincides with the set of $\chi_\lambda : \mathbb{R}^2 \longrightarrow \Lambda\mathbb{C}_0$ such that $\psi_\lambda \rightleftharpoons \chi_\lambda$.

Proof. We shall denote

$$
\begin{pmatrix} v_\lambda \\ w_\lambda \end{pmatrix} := \begin{pmatrix} \cosh\omega & -i\sinh\omega \\ i\sinh\omega & \cosh\omega \end{pmatrix} \begin{pmatrix} a_\lambda \\ b_\lambda \end{pmatrix}
$$

and

$$
\begin{pmatrix} \tilde{v}_\lambda \\ \tilde{w}_\lambda \end{pmatrix} := \begin{pmatrix} \cosh\omega & i\sinh\omega \\ -i\sinh\omega & \cosh\omega \end{pmatrix} \begin{pmatrix} a_\lambda \\ b_\lambda \end{pmatrix}.
$$

Step 1: Uniqueness of χ_λ modulo ω_z

Since the system (9.46) is linear in ψ_λ and χ_λ, it suffices to study the solutions of the system (9.46) for $\psi_\lambda = 0$. This reads

$$
\begin{cases}
\dfrac{\partial}{\partial \bar{z}} \begin{pmatrix} v_\lambda \\ w_\lambda \end{pmatrix} = 0 \\[4mm]
\dfrac{\partial}{\partial z} \begin{pmatrix} \cosh 2\omega & i\sinh 2\omega \\ -i\sinh 2\omega & \cosh 2\omega \end{pmatrix} \begin{pmatrix} v_\lambda \\ w_\lambda \end{pmatrix} = -\lambda^{-1}\chi_\lambda \begin{pmatrix} i\sinh 2\omega \\ \cosh 2\omega \end{pmatrix}
\end{cases}.
$$

The first equation means that v_λ and w_λ are holomorphic. The second equation gives

$$
\left(\frac{\partial w_\lambda}{\partial z} - 2i\omega_z v_\lambda \right) \begin{pmatrix} i\sinh 2\omega \\ \cosh 2\omega \end{pmatrix} + \left(\frac{\partial v_\lambda}{\partial z} + 2i\omega_z w_\lambda \right) \begin{pmatrix} \cosh 2\omega \\ -i\sinh 2\omega \end{pmatrix}
$$
$$
= -\lambda^{-1}\chi_\lambda \begin{pmatrix} i\sinh 2\omega \\ \cosh 2\omega \end{pmatrix}.
$$

Since $\left\{ \left(\begin{array}{c} \cosh 2\omega \\ -i\sinh 2\omega \end{array} \right), \left(\begin{array}{c} i\sinh 2\omega \\ \cosh 2\omega \end{array} \right) \right\}$ is a basis, this is equivalent to

$$
\left\{
\begin{array}{rcl}
\dfrac{\partial w_\lambda}{\partial z} - 2i\omega_z v_\lambda + \lambda^{-1}\chi_\lambda & = & 0 \\[2mm]
\dfrac{\partial v_\lambda}{\partial z} + 2i\omega_z w_\lambda & = & 0.
\end{array}
\right.
$$

Assume that $w_\lambda \neq 0$. Then since v_λ and w_λ are holomorphic, the second equation would imply that ω_z is meromorphic. This is however impossible (excepted in the case where $\omega \equiv 0$) because $(\omega_z)_{\bar{z}} = -\frac{1}{8}\sinh 2\omega \neq 0$. Thus we necessarily have $w_\lambda \equiv 0$ and the second relation implies $\frac{\partial v_\lambda}{\partial z} = 0$. Hence v_λ is equal to a constant $C_\lambda \in \Lambda\mathbb{C}$. Now the first equation gives $\chi_\lambda = 2i\lambda C_\lambda \omega_z$.

Step 2: Existence of a solution to system (9.46) as soon as $\psi_\lambda \rightleftharpoons \chi_\lambda$

We let ϕ_λ to be the solution of (9.45) such that $\chi_\lambda = 4\left(\lambda^2 \frac{\partial^2 \psi_\lambda}{(\partial z)^2} - \lambda \frac{\partial \omega}{\partial z}\phi_\lambda \right)$ and we set

$$
\left(\begin{array}{c} v_\lambda \\ w_\lambda \end{array} \right) = \left(\begin{array}{c} 2i\phi_\lambda \\ -4\lambda\frac{\partial \psi_\lambda}{\partial z} \end{array} \right).
$$

Then using (9.45) and (9.44) we obtain

$$
\frac{\partial}{\partial \bar{z}} \left(\begin{array}{c} v_\lambda \\ w_\lambda \end{array} \right) = \lambda\psi_\lambda \left(\begin{array}{c} -i\sinh 2\omega \\ \cosh 2\omega \end{array} \right).
$$

We then find that

$$
\frac{\partial}{\partial z} \left(\begin{array}{cc} \cosh 2\omega & i\sinh 2\omega \\ -i\sinh 2\omega & \cosh 2\omega \end{array} \right) \left(\begin{array}{c} v_\lambda \\ w_\lambda \end{array} \right)
$$
$$
= -4\lambda^{-1}\left(\lambda^2 \frac{\partial^2 \psi_\lambda}{(\partial z)^2} - \lambda\omega_z \phi_\lambda \right) \left(\begin{array}{c} i\sinh 2\omega \\ \cosh 2\omega \end{array} \right),
$$

and the conclusion follows. $\qquad\square$

Corollary 9.1 *If $\psi_\lambda : \mathbb{R}^2 \longrightarrow \Lambda\mathbb{C}_0$ and $\chi_\lambda : \mathbb{R}^2 \longrightarrow \Lambda\mathbb{C}_0$ are two solutions of (9.44), then $\psi_\lambda \rightleftharpoons \chi_\lambda$ if and only if $\overline{\chi_\lambda} \rightleftharpoons \overline{\psi_\lambda}$.*

Proof. The system (9.46) is invariant by complex conjugation provided one changes $(\psi_\lambda, \chi_\lambda)$ into $(-\overline{\chi_\lambda}, -\overline{\psi_\lambda})$ (and $(a_\lambda, b_\lambda) \mapsto (\overline{a_\lambda}, \overline{b_\lambda})$), so the result follows. $\qquad\square$

An algorithm to produce solutions to (9.44)

An obvious solution to (9.44) is ω_z. By Lemma 9.2 all maps χ_λ such that $\omega_z \rightleftharpoons \chi_\lambda$ are also solutions of (9.44). A computation shows that such χ_λ's are of the form

$$
4\lambda^2(\omega_{zzz} - 2\omega_z^3) + C_\lambda\omega_z.
$$

We may repeat this computation, finding that

$$4\lambda^2(\omega_{zzz} - 2\omega_z^3) \rightleftharpoons 4^2\lambda^4(\omega_{zzzzz} - 10\omega_{zzz}\omega_z^2 - 10\omega_{zz}^2\omega_z + 6\omega_z^5),$$

etc... We can hence generate a unique sequence $(\psi_\lambda^{(n)})_{l\in\mathbb{N}}$ defined by

- $\psi_\lambda^{(0)} = \lambda^{-2}P^{(0)}[\omega] := 0$
- $\psi_\lambda^{(1)} = P^{(1)}[\omega] := \omega_z$
- $\forall n \in \mathbb{N}, \psi_\lambda^{(n)} \rightleftharpoons \psi_\lambda^{(n+1)}$
- $\forall n \in \mathbb{N}, \psi_\lambda^{(n)} = \lambda^{2n-2}P^{(n)}[\omega]$ where $P^{(n)}[\omega]$ is a polynomial in $\mathbb{C}[\omega_z, \omega_{zz}, \ldots]$ homogeneous in $\frac{\partial}{\partial z}$ of degree $2n - 1$

The last assumption means that

$$P^{(n)}[\omega] = \sum_j \sum_{\alpha_1,\ldots,\alpha_j} \sum_j \sum_{\beta_1,\ldots,\beta_j}$$

$$C_n(\alpha_1,\ldots,\alpha_j,\beta_1,\ldots,\beta_j)\left(\frac{\partial^{\alpha_1}\omega}{(\partial z)^{\alpha_1}}\right)^{\beta_1}\cdots\left(\frac{\partial^{\alpha_j}\omega}{(\partial z)^{\alpha_j}}\right)^{\beta_j},$$

where $C_n(\alpha_1,\ldots,\alpha_j,\beta_1,\ldots,\beta_j) = 0$ if $\alpha_1\beta_1 + \cdots + \alpha_j\beta_j \neq 2n + 1$. It can be deduced from an algebraic recursion formula which is proved in [68], Proposition 3.1, p. 149.

An immediate consequence of Corollary 9.1 is that

$$\overline{\psi_\lambda^{(n+1)}} \rightleftharpoons \overline{\psi_\lambda^{(n)}}, \ \forall n \in \mathbb{N}.$$

Algebraic construction of a series of infinitesimal deformations in $T_{F_\lambda}\mathcal{E}$

We wish to construct families of solutions $\psi_\lambda : \mathbb{R}^2 \longrightarrow \Lambda\mathbb{R}_0$ of (9.44) such that there exist $\phi_\lambda : \mathbb{R}^2 \longrightarrow \Lambda\mathbb{C}$ solving (9.45) and such that $\Omega(\psi_\lambda, \phi_\lambda)$ is independent of λ.

By denoting $\chi_\lambda := 4\left(\lambda^2\frac{\partial^2\psi_\lambda}{(\partial z)^2} - \lambda\frac{\partial\omega}{\partial z}\phi_\lambda\right)$, this amounts to find pairs $(\psi_\lambda, \chi_\lambda)$ of solutions of (9.44) such that $\psi_\lambda \rightleftharpoons \chi_\lambda$ and $\frac{1}{4}\chi_\lambda + \frac{1}{4}\overline{\chi_\lambda} - \frac{1}{2}\psi_\lambda$ does not depend on λ.

We look for maps ψ_λ of the form

$$\psi_\lambda = \sum_{k=0}^n \lambda^{-2k}a_k P^{(n-k+1)}[\omega] + \sum_{k=0}^n \lambda^{2k}\overline{a_k}\overline{P^{(n-k+1)}[\omega]},$$

where $a_0, \ldots, a_n$ are complex constants to be determined. Note that this ansatz garantees that ψ_λ is real valued. The maps χ_λ such that $\psi_\lambda \rightleftharpoons \chi_\lambda$ are

$$\chi_\lambda = C_\lambda P^{(1)}[\omega] + \lambda^2 \sum_{k=0}^{n} \lambda^{-2k} a_k P^{(n-k+2)}[\omega] + \lambda^2 \sum_{k=0}^{n} \lambda^{2k} \overline{a_k} \overline{P^{(n-k)}[\omega]},$$

where $C_\lambda \in \Lambda\mathbb{C}$ and so

$$\frac{1}{4}\left(\chi_\lambda + \overline{\chi_\lambda} - 2\psi_\lambda\right)$$

$$= \frac{1}{4}\left[\sum_{k=0}^{n} \lambda^{-2k+2} a_k P^{(n-k+2)}[\omega] + \sum_{k=0}^{n-1} \lambda^{2k+2} \overline{a_k}\overline{P^{(n-k)}[\omega]} + C_\lambda P^{(1)}[\omega]\right.$$

$$\sum_{k=0}^{n} \lambda^{2k-2} \overline{a_k}\overline{P^{(n-k+2)}[\omega]} + \sum_{k=0}^{n-1} \lambda^{-2k-2} a_k P^{(n-k)}[\omega] + \overline{C_\lambda}\overline{P^{(1)}[\omega]}$$

$$\left.-2\sum_{k=0}^{n} \lambda^{-2k} a_k P^{(n-k+1)}[\omega] - 2\sum_{k=0}^{n} \lambda^{2k} \overline{a_k}\overline{P^{(n-k+1)}[\omega]}\right]$$

$$= \frac{1}{4}\left[\lambda^{-2n}(a_{n-1} - 2a_n)P^{(1)}[\omega] + C_\lambda P^{(1)}[\omega]\right.$$
$$+\lambda^{-2n+2}(a_n + a_{n-2} - 2a_{n-1})P^{(2)}[\omega]$$
$$+\cdots$$
$$+\lambda^{-2}(a_2 + a_0 - 2a_1)P^{(n)}[\omega] + \lambda^{-2}\overline{a_0}\overline{P^{(n+2)}[\omega]}$$
$$+(a_1 - 2a_0)P^{(n+1)}[\omega] + (\overline{a_1} - 2\overline{a_0})\overline{P^{(n+1)}[\omega]}$$
$$+\lambda^2(\overline{a_2} + \overline{a_0} - 2\overline{a_1})\overline{P^{(n)}[\omega]} + \lambda^2 a_0 P^{(n+2)}[\omega]$$
$$+\cdots$$
$$+\lambda^{2n-2}(\overline{a_{n-2}} + \overline{a_n} - 2\overline{a_{n-1}})\overline{P^{(2)}[\omega]}$$
$$\left.+ \lambda^{2n}(\overline{a_{n-1}} - 2\overline{a_n})\overline{P^{(1)}[\omega]} + \overline{C_\lambda}\overline{P^{(1)}[\omega]}\right].$$

It turns out that this quantity is independent of λ if and only if we choose

$$a_k = kb, \ \forall k \in \mathbb{N} \ \text{ and } \ C_\lambda = (n+1)b\lambda^{-2n} + c$$

for some constants $b, c \in \mathbb{C}$. We will choose $c = 0$. Then we are left with [5]

$$\Omega(\psi_\lambda, \phi_\lambda) = \frac{1}{4}\left(\chi_\lambda + \overline{\chi_\lambda} - 2\psi_\lambda\right) = \frac{1}{4}\left(bP^{(n+1)}[\omega] + \overline{b}\overline{P^{(n+1)}[\omega]}\right).$$

---

[5]in general for $c \neq 0$,

$$\frac{1}{4}\left(\chi_\lambda + \overline{\chi_\lambda} - 2\psi_\lambda\right) = \frac{1}{4}\left(bP^{(n+1)}[\omega] + \overline{b}\overline{P^{(n+1)}[\omega]} + cP^{(1)}[\omega] + \overline{c}\overline{P^{(1)}[\omega]}\right)$$

We conclude: for any $n \in \mathbb{N}$ and $b \in \mathbb{C}$ we define the maps

$$\Psi_\lambda^{(n,b)} := b \sum_{k=1}^{n} k\lambda^{-2k} P^{(n-k+1)}[\omega] + \bar{b} \sum_{k=1}^{n} k\lambda^{2k} \overline{P^{(n-k+1)}[\omega]}.$$

Then there exists a unique map $\Phi_\lambda^{(n,b)}$ such that (9.45) holds and such that

$$\Omega(\Psi_\lambda^{(n,b)}, \Phi_\lambda^{(n,b)}) = \frac{1}{4}\left(bP^{(n+1)}[\omega] + \bar{b}\overline{P^{(n+1)}[\omega]}\right)$$

and so in particular is independent of λ. By Theorem 9.2 there exists a unique $T_\lambda^{(n,b)} \in T_{F_\lambda}\mathcal{E}$ such that $t_\lambda^{(n,b),3} = \Psi_\lambda^{(n,b)}$ and $\dot{\omega}^{(n,b)} = \Omega(\Psi_\lambda^{(n,b)}, \Phi_\lambda^{(n,b)})$. $T_\lambda^{(n,b)}$ is defined by the formulas in Lemma 9.1.

Examples
a) For $\Psi_\lambda^{(0,b)} = 0$, where $b \in \mathbb{C}$, we have

$$\Phi_\lambda^{(0,b)} = -\frac{b}{4}\lambda^{-1}, \quad \chi_\lambda^{(0,b)} = ib\omega_z$$

and

$$\dot{\omega}^{(0,b)} = \Omega(\Psi_\lambda^{(0,b)}, \Phi_\lambda^{(0,b)}) = \frac{1}{4}\left(b\omega_z + \bar{b}\overline{\omega_z}\right),$$

$$t_{2,\lambda}^{(0,b),1} = \frac{i}{4}\left(b\omega_z - \bar{b}\overline{\omega_z}\right).$$

b) For

$$\Psi_\lambda^{(1,b)} = \lambda^{-2}b\omega_z + \lambda^2 \bar{b}\omega_{\bar{z}},$$

where $b \in \mathbb{C}$, we have

$$\Phi_\lambda^{(1,b)} = -\frac{b}{2}\lambda^{-3} + 2\lambda^{-1}b\omega_z^2 - \lambda^3\frac{\bar{b}}{4}\cosh 2\omega,$$

$$\chi_\lambda^{(1,b)} = 2b\lambda^{-2}\omega_z + 4b(\omega_{zzz} - \omega_z^3),$$

and

$$\dot{\omega}^{(1,b)} = \Omega(\Psi_\lambda^{(1,b)}, \Phi_\lambda^{(1,b)}) = b(\omega_{zzz} - \omega_z^3) + \bar{b}\overline{(\omega_{zzz} - \omega_z^3)},$$

$$t_{2,\lambda}^{(1,b),1} = \frac{ib}{2}\lambda^{-2}\omega_z + ib(\omega_{zzz} - \omega_z^3) - i\bar{b}\overline{(\omega_{zzz} - \omega_z^3)} - \frac{i\bar{b}}{2}\lambda^2\omega_{\bar{z}}.$$

10 Wente tori

Historically the first immersed constant mean curvature tori were constructed by H. Wente in 1984 [86] by analytical methods. A short time later U. Abresch simplified this construction. He remarked that Wente tori should possess planar curvature lines and thus studied all CMC surfaces with planar curvature lines. It leads to an overdetermined system of equations which can be solved by quadratures using elliptic integrals. And U. Abresch showed that some of the obtained immersed surfaces do close up, giving CMC tori [1] (see also [87]).

A similar treatment was given in [84], [85] by R. Walter who remarked that each curvature line of these tori is either planar or spherical. Moreover, as pointed out in [66] special surfaces (including constant mean curvature ones) with planar or spherical curvature lines were already investigated by A. Enneper [37], [38], [39] and his students [63], [15], [83], (see also [27]), [28], [61], [3], in the last century. Explicit formulas were then produced.

10.1 CMC surfaces with planar curvature lines

We again look at conformal CMC immersions $X : \mathbb{R}^2 \longrightarrow \mathbb{R}^3$ such that $H = 1/2$ and such that their Gauss map $u : \mathbb{R}^2 \longrightarrow S^2$ satisfies

$$f = \left| \frac{\partial u}{\partial x} \right|^2 - \left| \frac{\partial u}{\partial y} \right|^2 - 2i \left\langle \frac{\partial u}{\partial x}, \frac{\partial u}{\partial y} \right\rangle = -1.$$

Then the curvature lines are the image of the curves $[x \longmapsto X(x,y)]$ (small curvature lines) and $[y \longmapsto X(x,y)]$ (large curvature lines). Let us focus at small curvature lines as an instance. Their Frenet framing $(\vec{t}_1, \vec{n}_1, \vec{b}_1)$ is defined by

$$
\begin{aligned}
\vec{t}_1 &= e^{-\omega} \frac{\partial X}{\partial x} \\
\vec{n}_1 &= \left| \frac{\partial \vec{t}_1}{\partial x} \right|^{-1} \frac{\partial \vec{t}_1}{\partial x} \\
\vec{b}_1 &= \vec{t}_1 \times \vec{n}_1.
\end{aligned}
$$

The curvature and the torsion along these lines are defined by

$$\kappa_1 := e^{-\omega} \left| \frac{\partial \vec{t}_1}{\partial x} \right|, \quad \tau_1 := -e^{-\omega} \left\langle \frac{\partial \vec{b}_1}{\partial x}, \vec{n}_1 \right\rangle.$$

Computing these datas, one finds that

$$\vec{t_1} \;=\; e_1$$

$$\kappa_1 \;=\; e^{-\omega}\sqrt{\left(\frac{\partial\omega}{\partial y}\right)^2 + \sinh^2\omega}$$

$$\kappa_1\vec{n_1} \;=\; e^{-\omega}\left(-\frac{\partial\omega}{\partial y}e_2 + \sinh\omega e_3\right)$$

$$\kappa_1\vec{b_1} \;=\; e^{-\omega}\left(-\sinh\omega e_2 - \frac{\partial\omega}{\partial y}e_3\right)$$

$$\kappa_1^2\tau_1 \;=\; e^{-3\omega}\left(\sinh\omega\frac{\partial^2\omega}{\partial x\partial y} - \cosh\omega\frac{\partial\omega}{\partial x}\frac{\partial\omega}{\partial y}\right).$$

The small curvature lines are planar if and only if the torsion τ_1 vanishes identically. This is clearly equivalent to the condition

$$\sinh\omega\frac{\partial^2\omega}{\partial x\partial y} - \cosh\omega\frac{\partial\omega}{\partial x}\frac{\partial\omega}{\partial y} = 0. \tag{10.1}$$

Similarly the large curvature lines are planar if and only if

$$\cosh\omega\frac{\partial^2\omega}{\partial x\partial y} - \sinh\omega\frac{\partial\omega}{\partial x}\frac{\partial\omega}{\partial y} = 0. \tag{10.2}$$

U. Abresch showed that choosing one the two above conditions leads to an overdetermined system of equations which can be solved explicitly using elliptic integrals. This is based on the observation that for instance (10.1) can be restated as

$$\frac{\partial}{\partial x}\left(\frac{1}{\sinh\omega}\frac{\partial\omega}{\partial y}\right) = \frac{\partial}{\partial y}\left(\frac{1}{\sinh\omega}\frac{\partial\omega}{\partial x}\right) = 0,$$

which implies that one may find two functions f and g of one real variable such that

$$\frac{1}{\sinh\omega}\frac{\partial\omega}{\partial x} = -f(x), \qquad \frac{1}{\sinh\omega}\frac{\partial\omega}{\partial y} = -g(y).$$

Thus the geometrical hypothesis (10.1) leads actually to a separation of variables. Using these new variables together with the equation

$$\Delta\omega + \cosh\omega\sinh\omega = 0,$$

one can then show by a tedious computation that f and g are solutions of

$$(f')^2 = f^4 + (1 + \alpha^2 - \beta^2)f^2 + \alpha^2,$$
$$(g')^2 = g^4 + (1 - \alpha^2 + \beta^2)g^2 + \beta^2,$$

for some constants α and β. Thus f and g can be given using elliptic integrals. A similar conclusion holds if one assumes (10.2) instead of (10.1). However only CMC immersions with planar *small curvature lines* can close up into immersed compact tori.

10.2 A system of commuting ordinary equations

In a subsequent work [2], U. Abresch proved that the above solutions arise from commuting flows. Namely we consider $\mathbb{R}^6$ with the coordinates

$$U := \begin{pmatrix} c \\ s \\ u_1 \\ u_2 \\ u_3 \\ u_4 \end{pmatrix},$$

and we define on $\mathbb{R}^6$ the two vector fields

$$X_1(U) := \frac{\sqrt{2}}{2} \begin{pmatrix} su_1 \\ cu_1 \\ su_3 - cs \\ cu_4 \\ -su_1 - u_2 u_4 \\ -cu_2 + u_2 u_3 \end{pmatrix}, \quad X_2(U) := \frac{\sqrt{2}}{2} \begin{pmatrix} su_2 \\ cu_2 \\ cu_4 \\ -su_3 - cs \\ su_2 + u_1 u_4 \\ -cu_1 - u_1 u_3 \end{pmatrix}.$$

One can check that X_1 and X_3 commute by a direct computation. Moreover the quantities $c^2 - s^2$ and $s^2 + u_1^2 + u_2^2 + u_3^2 + u_4^2$ are preserved by X_1 and X_2 so their flows remain in a compact subset of $\mathbb{R}^6$ and thus are defined for all time. So let us take any solution $U : \mathbb{R}^2 \longrightarrow \mathbb{R}^6$ of

$$\frac{\partial U}{\partial x} = X_1(U), \quad \text{and} \quad \frac{\partial U}{\partial y} = X_2(U) \tag{10.3}$$

with initial conditions (at $(x, y) = 0$) such that

$$c^2(0) - s^2(0) = 1. \tag{10.4}$$

Then since the quantity $c^2 - s^2$ is preserved by the flow, we can pose $c = \cosh \omega$ and $s = \sinh \omega$, where ω is a function on $\mathbb{R}^2$. Equations (10.3) can then be rewritten as

$$\frac{\partial \omega}{\partial z} = \frac{1}{2\sqrt{2}} (u_1 - iu_2), \tag{10.5}$$

$$\frac{\partial^2 \omega}{(\partial z)^2} = \frac{1}{4} (\sinh \omega u_3 - i \cosh \omega u_4), \tag{10.6}$$

$$\frac{\partial^2 \omega}{\partial z \partial \bar{z}} = -\frac{1}{4} \cosh \omega \sinh \omega. \tag{10.7}$$

We see hence that ω is a solution of the sinh-Gordon equation. So we can associate conformal CMC immersions to each solution of (10.3) with the initial condition (10.4). These immersions turn to enjoy two properties:

- they contains all conformal CMC immersions with planar small curvature lines or planar large curvature lines (see [2]) and in particular "Wente" tori (actually (10.1) is equivalent in choosing $a_2 = 0$ in (10.9) below)
- they are of finite type, as we will see in the next paragraph

10.3   Recovering a finite type solution

We analyze the conformal CMC immersions obtained by integrating (10.3) with the initial condition (10.4). We can first observe (as in [2]) that the following quantities are conserved along the flow of X_1 and X_2:

$$a_1 := cu_3 - \frac{1}{2}(u_1^2 - u_2^2),\qquad\qquad (10.8)$$

$$a_2 := su_4 - u_1 u_2.\qquad\qquad (10.9)$$

This is proved by a straightforward computation. We take a solution of (10.3) with $c^2 - s^2 = 1$, so that we may pose $c = \cosh\omega$ and $s = \sinh\omega$. Then all higher derivatives of ω can be expressed in terms of algebraic functions of U: the first relations are precisely (10.5), (10.6) and (10.7). The next one consists in computing $\frac{\partial^3\omega}{(\partial z)^3}$. Using first (10.6) and (10.3)

$$\begin{aligned}
\frac{\partial^3\omega}{(\partial z)^3} &= \frac{1}{8}\left(\frac{\partial}{\partial x} - i\frac{\partial}{\partial y}\right)(su_3 - icu_4) \\
&= \frac{1}{8\sqrt{2}}(u_1 - iu_2)(1 + 2cu_3 - i2su_4).
\end{aligned}$$

And now, by (10.8) and (10.9),

$$\begin{aligned}
\frac{\partial^3\omega}{(\partial z)^3} &= \frac{1}{8\sqrt{2}}(u_1 - iu_2)(1 + 2a_1 - i2a_2 + (u_1 - iu_2)^2) \\
&= \frac{1 + 2a_1 - i2a_2}{4}\frac{\partial\omega}{\partial z} + 2\left(\frac{\partial\omega}{\partial z}\right)^3,
\end{aligned}$$

where we have used (10.5) in the last line. We thus conclude that

$$4\lambda^2\left(\frac{\partial^3\omega}{(\partial z)^3} - 2\left(\frac{\partial\omega}{\partial z}\right)^3\right) = \lambda^2(1 + 2a_1 - i2a_2)\frac{\partial\omega}{\partial z},$$

i. e., with the notations of Chapter 9, $\omega_z \rightleftharpoons \lambda^2(1 + 2a_1 - i2a_2)\omega_z$, which is equivalent to $\omega_z \rightleftharpoons 0$. In the light of the analysis of Chapter 9, it just means that ω is a solution of finite type. In particular the relevant deformation $T_\lambda^\star$ of the surface (such that property (9.9) holds) can be chosen so that

$$t_\lambda^{\star,3} = \lambda^{-2}\omega_z + \lambda^2\omega_{\bar{z}}.$$

10.4 Spectral curves

Any finite type solution is characterized by the existence of a map $\eta_\lambda : \mathbb{R}^2 \longrightarrow \Lambda^d \mathfrak{g}_\tau$ which is a stationary solution of (8.4): $d\eta_\lambda + [\alpha_\lambda, \eta_\lambda] = 0$. Assuming that η_λ and α_λ are matrices we can make sense of the characteristic determinant

$$P(\lambda, \mu) := \det(\eta_\lambda - \mu \mathbb{1}).$$

Then equation (8.4) just implies that this polynomial does not depend on z. By compactifying the complex algebraic curve of $\mathbb{C}^2$ of equation $P(\lambda, \mu) = 0$, we obtain a Riemann surface called the *spectral curve* of the finite type solution. Lastly it is possible to use tools from algebraic geometry (the Abel-Jacobi integrals and the Jacobi torus of the spectral curve, see [32], [19]) to produce expressions of the conformal CMC immersions in terms of theta functions. This has been carried out by A. Bobenko for surfaces in $\mathbb{R}^3$, S^3 and $\mathbb{H}^3$ [11], [12] (see also [13]).

In particular Wente tori correspond to genus 2 spectral curves invariant by a complex involution [13].

11 Weierstrass type representations

The theory for finite type solutions developped in Chapter 8 can be generalized in order to represent all harmonic maps from a simply connected surface to symmetric spaces like the sphere S^2. This has been developped by J. Dorfmeister, F. Pedit and H.Y. Wu and leads to a Weierstrass type representation [30].

The main idea is to try to generalize property (8.5) $F_\lambda B_\lambda = e^{\lambda^{d-1}\overset{\circ}{\eta}_\lambda z}$ which was found for all finite type solutions (see paragraph 8.4), by substituting the right hand side of (8.5) by a suitable holomorphic function of z with values in a complexified loop group. One then needs to find a way to split this holomorphic function into a product of the form $F_\lambda B_\lambda$, i.e. we need a non linear analog of the Lie algebra splitting (8.1). Such a splitting does exist and can actually be interpreted as an extension to loop groups of the Iwasawa decomposition described in paragraph 8.1.

We will present here briefly this theory. We will be concerned with harmonic maps u from a simply connected domain $\Omega \subset \mathbb{R}^2$ into a symmetric manifold $\mathfrak{G}/\mathfrak{K}$. We assume that there exists an automorphism $\tau : \mathfrak{G} \longrightarrow \mathfrak{G}$ which is an involution ($\tau^2 = \mathbb{1}$) such that, if $\mathfrak{G}_\tau := \{g \in \mathfrak{G} : \tau(g) = g\}$, $(\mathfrak{G}_\tau)_0 \subset \mathfrak{K} \subset \mathfrak{G}_\tau$.

11.1 Loop groups decompositions

We use here the notations $\Lambda\mathfrak{G}_\tau$, $\Lambda\mathfrak{G}_\tau^{\mathbb{C}}$ concerning loop groups and twisted loop groups introduced in paragraph 7.2. We also will need the loop group $\Lambda_{\mathfrak{B}_\mathfrak{K}}^+\mathfrak{G}_\tau^{\mathbb{C}}$ defined in paragraph 8.4. We assume that $\mathfrak{K}$ is a compact semi-simple group, so that the following Iwasawa decomposition holds:

$$
\begin{array}{ccc}
\mathfrak{K} \times \mathfrak{B}_\mathfrak{K} & \longrightarrow & \mathfrak{K}^{\mathbb{C}} \\
(g, b) & \longmapsto & gb
\end{array}
$$

is a diffeomorphism. We summarize by $\mathfrak{K}^{\mathbb{C}} = \mathfrak{K}.\mathfrak{B}_\mathfrak{K}$ this property. Then the first tool is an extension of this decomposition to the loop group $\Lambda\mathfrak{G}_\tau^{\mathbb{C}}$, namely

Lemma 11.1 *Assume that $\mathfrak{G}$ is a compact semisimple Lie group. Then the mapping*

$$
\begin{array}{ccc}
\Lambda\mathfrak{G}_\tau \times \Lambda_{\mathfrak{B}_\mathfrak{K}}^+\mathfrak{G}_\tau^{\mathbb{C}} & \longrightarrow & \Lambda\mathfrak{G}_\tau^{\mathbb{C}} \\
(g_\lambda, b_\lambda) & \longmapsto & g_\lambda b_\lambda
\end{array}
$$

is a diffeomorphism. We denote by $\Lambda\mathfrak{G}_\tau^{\mathbb{C}} = \Lambda\mathfrak{G}_\tau.\Lambda_{\mathfrak{B}_\mathfrak{K}}^+\mathfrak{G}_\tau^{\mathbb{C}}$ this property.

We also introduce the subgroups

$$\Lambda^+\mathfrak{G}_\tau^{\mathbb{C}} := \{[\lambda \mapsto \phi_\lambda] \in \Lambda\mathfrak{G}_\tau^{\mathbb{C}} \text{ extending holomorphically in the disk } D^2\},$$

$$\Lambda_\star^-\mathfrak{G}_\tau^{\mathbb{C}} := \{[\lambda \mapsto \phi_\lambda] \in \Lambda\mathfrak{G}_\tau^{\mathbb{C}} \text{ extending holomorphically}$$
$$\text{in } \mathbb{C} \cup \{\infty\}\backslash D^2 \text{ and } \phi_\infty = 1\}.$$

We then have the following.

Lemma 11.2 *Assume that $\mathfrak{G}$ is a semisimple Lie group. Then there exists a dense open subset $\mathcal{C}$ of the connected component of the identity of $\Lambda\mathfrak{G}_\tau^{\mathbb{C}}$, called the* big cell, *such that the mapping*

$$\begin{array}{ccc} \Lambda_\star^-\mathfrak{G}_\tau^{\mathbb{C}} \times \Lambda^+\mathfrak{G}_\tau^{\mathbb{C}} & \longrightarrow & \mathcal{C} \\ (\phi_\lambda^-, \phi_\lambda^+) & \longmapsto & \phi_\lambda^-\phi_\lambda^+ \end{array}$$

is a diffeomorphism. We denote by $\mathcal{C} = \Lambda_\star^-\mathfrak{G}_\tau^{\mathbb{C}}.\Lambda^+\mathfrak{G}_\tau^{\mathbb{C}}$ this property.

These two results are showed in [30]. The proofs are based on similar decomposition results for non twisted loop groups proved in [70].

11.2 Solutions in terms of holomorphic data

Extended lifts of harmonic maps are in correspondance with holomorphic datas as defined below. We first denote

$$\Lambda_{-1,\infty}\mathfrak{g}_\tau^{\mathbb{C}} := \{[\lambda \mapsto \xi_\lambda] \in \Lambda\mathfrak{g}_\tau^{\mathbb{C}}/\xi_\lambda = \sum_{k=-1}^{\infty} \hat{\xi}_k\lambda^k\}.$$

Definition 11.1 *The set of* holomorphic potentials, *denoted $\mathcal{H}_{-1,\infty}(\Omega)$, is the set of holomorphic 1-forms on Ω with values in $\Lambda_{-1,\infty}\mathfrak{g}_\tau^{\mathbb{C}}$. So any form μ_λ in $\mathcal{H}_{-1,\infty}(\Omega)$ has the expression*

$$\mu_\lambda = \sum_{k=-1}^{\infty} \hat{\mu}_k\lambda^k = \sum_{k=-1}^{\infty} \hat{\xi}_k(z)\lambda^k dz,$$

where $\forall z \in \Omega, \sum_{k=-1}^{\infty} \hat{\xi}_k(z)\lambda^k \in \Lambda_{-1,\infty}\mathfrak{g}_\tau^{\mathbb{C}}$.

Lemma 11.3 *Let $F_\lambda : \Omega \to \Lambda\mathfrak{G}_\tau$ be the extended lift of a harmonic map $u : \Omega \longrightarrow \mathfrak{G}/\mathfrak{K}$ and assume that Ω is contractible. Then*

- *there exist a holomorphic map $H_\lambda : \Omega \to \Lambda\mathfrak{G}_\tau^{\mathbb{C}}$ and a map $B_\lambda : \Omega \to \Lambda_{\mathfrak{B}_\mathfrak{K}}^+\mathfrak{G}_\tau^{\mathbb{C}}$ such that $F_\lambda = H_\lambda B_\lambda$.*
- *the Maurer-Cartan form $\mu_\lambda := (H_\lambda)^{-1}dH_\lambda$ is a holomorphic potential.*

Proof. (see [30] for details) The existence of H_λ and B_λ relies on solving the equation

$$0 = \frac{\partial(F_\lambda B_\lambda^{-1})}{\partial \bar{z}} = F_\lambda \left(\alpha_\lambda \left(\frac{\partial}{\partial \bar{z}} \right) - B_\lambda^{-1} \frac{\partial B_\lambda}{\partial \bar{z}} \right) (B_\lambda)^{-1},$$

which is equivalent to

$$\frac{\partial B_\lambda}{\partial \bar{z}} = B_\lambda (\alpha_0 + \lambda \alpha_1 + \lambda^2 \alpha_2) \left(\frac{\partial}{\partial \bar{z}} \right),$$

with the constraint that B_λ takes values in $\Lambda^+_{\mathfrak{B}_\mathfrak{K}} \mathfrak{G}^{\mathbb{C}}_\tau$. The existence of a solution is first obtained locally, then we can glue local solutions into a global one. This proves the first assertion. Now we write

$$(H_\lambda)^{-1} dH_\lambda = B_\lambda (\alpha_\lambda - B_\lambda^{-1} dB_\lambda) B_\lambda^{-1},$$

and using the fact that B_λ takes values in $\Lambda^+_{\mathfrak{B}_\mathfrak{K}} \mathfrak{G}^{\mathbb{C}}_\tau$ and that $z \mapsto H_\lambda(z)$ is holomorphic, we deduce that $\mu_\lambda := H_\lambda^{-1} dH_\lambda$ has the desired properties. $\quad\square$

Conversely any holomorphic potential in $\mathcal{H}_{-1,\infty}(\Omega)$ produces harmonic maps as follows.

Theorem 11.1 *Assume that $\mathfrak{G}$ is compact semi-simple. Let $\mu_\lambda \in \mathcal{H}_{-1,\infty}(\Omega)$, p_0 a point in Ω and H_λ^0 a constant in $\Lambda \mathfrak{G}^{\mathbb{C}}_\tau$. Then*

- *there exists a unique holomorphic map $H_\lambda : \Omega \to \Lambda \mathfrak{G}^{\mathbb{C}}_\tau$, such that $dH_\lambda = H_\lambda \mu_\lambda$ and $H_\lambda(p_0) = H_\lambda^0$.*
- *we can apply the loop groups decomposition $\Lambda \mathfrak{G}^{\mathbb{C}}_\tau = \Lambda \mathfrak{G}_\tau . \Lambda^+_{\mathfrak{B}_\mathfrak{K}} \mathfrak{G}^{\mathbb{C}}_\tau$ to $H_\lambda(z)$ for all value of z. It follows that there exist two maps $F_\lambda : \Omega \to \Lambda \mathfrak{G}_\tau$ and $B_\lambda : \Omega \to \Lambda^+_{\mathfrak{B}_{\mathfrak{G}_0}} \mathfrak{G}^{\mathbb{C}}_\tau$ such that*

$$H_\lambda(z) = F_\lambda(z) B_\lambda(z), \quad \forall z \in \Omega.$$

Then F_λ is an extended lift of a harmonic map $\Omega \longrightarrow \mathfrak{G}/\mathfrak{K}$.

Proof. Since $\mu_\lambda = \xi_\lambda dz$, with $\frac{\partial \xi_\lambda}{\partial \bar{z}} = 0$, it follows easily that $d\mu_\lambda + \mu_\lambda \wedge \mu_\lambda = 0$, hence the existence and the uniqueness of H_λ. Using now Lemma 11.1 in order to perform the decomposition $H_\lambda = F_\lambda B_\lambda$, we obtain

$$(F_\lambda)^{-1} dF_\lambda = B_\lambda \mu_\lambda (B_\lambda)^{-1} - dB_\lambda (B_\lambda)^{-1}. \tag{11.1}$$

Now using the fact that $\mu_\lambda \in \mathcal{H}_{-1,\infty}(\Omega)$ and B_λ takes value in $\Lambda^+_{\mathfrak{B}_\mathfrak{K}} \mathfrak{G}^{\mathbb{C}}_\tau$, it is easy to check that the right hand side of (11.1) has the form $\sum_{k=-1}^{\infty} \hat{\alpha}_k \lambda^k$. But (11.1) implies also that this quantity should be real, i.e. a 1-form with coefficients in $\Lambda \mathfrak{G}_\tau$. Hence $\alpha_\lambda := F_\lambda^{-1} dF_\lambda$ reduces to $\alpha_\lambda = \hat{\alpha}_{-1} \lambda^{-1} + \hat{\alpha}_0 + \hat{\alpha}_1 \lambda$ and moreover $\hat{\alpha}_0$ is real and $\hat{\alpha}_1 = \bar{\hat{\alpha}}_{-1}$. Lastly a Taylor expansion in λ of (11.1) proves that $\hat{\alpha}_{-1}$ is a (1,0)-form, which ensures the result by Chapter 7. $\quad\square$

11.3   Meromorphic potentials

The holomorphic potentials constructed in Lemma 11.3 are far from being unique. Moreover they involve in general infinitely many holomorphic maps. These defects can be mended, provided we allow meromorphic potentials and under some hypotheses on $\mathfrak{G}$. We define

$$\Lambda_{-1}\mathfrak{g}_\tau^{\mathbb{C}} := \{[\lambda \mapsto \phi_\lambda] \in \Lambda\mathfrak{g}_\tau^{\mathbb{C}}/\xi_\lambda = \hat{\xi}_{-1}\lambda^{-1}\}.$$

Definition 11.2 *The set of meromorphic potentials, denoted $\mathcal{M}_{-1}(\Omega)$, is the set of meromorphic 1-forms on Ω with coefficients in $\Lambda_{-1}\mathfrak{g}_\tau^{\mathbb{C}}$. So any form μ_λ in $\mathcal{M}_{-1}(\Omega)$ has the expression*

$$\mu_\lambda = \hat{\mu}_{-1}\lambda^{-1} = \hat{\xi}_{-1}\lambda^{-1}(z)dz,$$

where $\hat{\xi}_{-1}(z)\lambda^{-1} \in \Lambda_{-1,\infty}\mathfrak{g}_\tau^{\mathbb{C}}$.

Then using the same methods as in [30]; one can prove the following

Theorem 11.2 *Assume that $\mathfrak{G}$ is semi-simple. Let $F_\lambda : \Omega \to \Lambda\mathfrak{G}_\tau$ be the extended lift of a harmonic map into $\mathfrak{G}/\mathfrak{K}$. Then there exists a finite subset $\{a_1, \ldots, a_p\}$ of Ω such that*

- *there exists a holomorphic map $F_\lambda^- : \Omega \setminus \{a_1, \ldots, a_p\} \to \Lambda_*^-\mathfrak{G}_\tau^{\mathbb{C}}$ and a map $F_\lambda^+ : \Omega \setminus \{a_1, \ldots, a_p\} \to \Lambda^+\mathfrak{G}_\tau^{\mathbb{C}}$ such that*

$$F_\lambda(z) = F_\lambda^-(z)F_\lambda^+(z), \quad \forall z \in \Omega \setminus \{a_1, \ldots, a_p\}$$

- *$z \mapsto F_\lambda^-(z)$ extends to a meromorphic map on Ω*
- *the Maurer-Cartan form $\mu_\lambda := (F_\lambda^-)^{-1}dF_\lambda^-$ of F_λ^- is a meromorphic potential in $\mathcal{M}_{-1}(\Omega)$.*

Proof. (see [30] for details) The decomposition $F_\lambda(z) = F_\lambda^-(z)F_\lambda^+(z)$ is possible as soon as we can prove that $F_\lambda(z)$ belongs to the big cell $\mathcal{C}$. Using Lemma 11.3 in the same way as in [30], one can show that this is true for all z, excepted maybe on a finite subset $\{a_1, \ldots, a_p\} \subset \Omega$. The second property is proved also in [30]. The last one follows easily by writing

$$\mu_\lambda = F_\lambda^+ \left[\alpha_\lambda - (F_\lambda^+)^{-1}dF_\lambda^+\right](F_\lambda^+)^{-1}$$

which implies on one hand that μ_λ is in $\mathcal{H}_{-1,\infty}(\Omega \setminus \{a_1, \ldots, a_p\})$, once one keep in mind the fact that $F_\lambda^+(z) \in \Lambda^+\mathfrak{G}_\tau^{\mathbb{C}}$. But on the other hand $F_\lambda^-(z) \in \Lambda_*^-\mathfrak{G}_\tau^{\mathbb{C}}$ and thus there is no nonnegative power of λ in the Fourier expansion of μ_λ. This implies the conclusion. $\qquad\square$

11.4   Generalizations

This theory can be extended to other two-dimensional differential geometric situations. An instance is the Willmore surface problem [88]: it consists in looking at surfaces Σ immersed in $\mathbb{R}^3$ which are critical points of the functional $\mathcal{W}[\Sigma] := \int_\Sigma H^2 dA$, where H is the mean curvature along Σ and dA is the area element induced by the first fundamental form of the immersion. These critical points are named *Willmore surfaces* and they satisfy the order four Euler-Lagrange equation $\Delta_\Sigma H + 2H(H^2 - K) = 0$. This problem was actually considered at the beginning of the century by G. Thomsen (se [81], [9]). Willmore surfaces may be thought as an analog of the minimal surface problem in conformal geometry and turns to be completely integrable. A Weierstrass representation theory in the spirit of the above theory was constructed in [49] (see also [50]).

Another instance is the study of Hamiltonian stationary Lagrangian surfaces in a four-dimensional Kähler manifold. They are Lagrangian surfaces which are critical points of the area functional under Hamiltonian flow deformations. The Lagrangian constraint and the restriction on the testing deformation vector fields lead to an order three Euler-Lagrange equation. In cases where the ambient space is a symmetric Kähler manifold, it can be shown that this is a completely integrable system and a Weierstrass type representation can also be built [51], [52], [53].

Bibliography

[1] U. Abresch, *Constant mean curvature tori in terms of elliptic functions*, J. reine angew. Math. 374 (1987), p. 169–192.

[2] U. Abresch, *Old and new periodic solutions of the sinh-Gordon equations*, Seminar on new results in non-linear partial differential equations, Vieweg, Wiesbaden 1987.

[3] P. Adam, *Sur les surfaces isothermiques à lignes de courbure planes dans un système ou dans les deux systèmes*, Ann. Sci. Ec. Norm. Sup. 10 (3) (1893), p. 319–358.

[4] M. Adler, P. van Moerbeke, *Completely integrable systems, Euclidean Lie algebras and curves*, Adv. Math. 38 (1980), p. 267–317.

[5] A.D. Aleksandrov, *Uniqueness theorems for surfaces in the large*, V. Amer. Math. Soc. Transl. 21 (1962), p. 412–416.

[6] M.V. Babisch, *Willmore surfaces, 4-particles Toda lattice and double covering of hyperelliptic surfaces*, Amer. Math. Soc. Transl. (2), 174, AMS (1996), p. 143–168.

[7] M.V. Babisch, A.I. Bobenko, *Willmore tori with umbilic lines and minimal surfaces in hyperbolic space*, Duke Math. J. 72 (1993), p. 151–185.

[8] L. Bianchi, *Lezioni di geometria differenziale*, Spoerri, Pisa, 1902.

[9] W. Blaschke, *Vorlesungen über Differentialgeometrie III*, Springer Berlin 1929.

[10] A.I. Bobenko, *Integrable surfaces*, Funkts. Anal. Prilozh. 24 (3) (1990), p. 68–69.

[11] A.I. Bobenko, *All constant mean curvature tori in $\mathbb{R}^3$, $\mathbb{S}^3$, $\mathbb{H}^3$ in terms of theta-functions*, Math. Ann. 290 (1991), p. 209–245.

[12] A.I. Bobenko, *Surfaces in terms of 2 by 2 matrices. Old and new integrable cases*, in [43].

[13] A.I. Bobenko, *Exploring surfaces through methods from the theory of integrable systems. Lectures on the Bonnet problem*, Lectures given on 12–30 April 1999 at the School on Differential Geometry at the ICTP, Trieste, preprint Technische Universität Berlin.

[14] A. Bobenko, U. Pinkall, *Discrete isothermic surfaces*, J. reine angew. Math. 475 (1996), p. 187–208.

[15] G. Bockwoldt, *Über die Enneperschen Flächen mit konstanten positivem Krümmungsmass*, Dissertation, Universität Göttingen, 1878.

[16] J. Bolton, L. Woodward, *The affine Toda equations and minimal surfaces*, in [43]

[17] O. Bonnet, *Note sur une propriété de maximum relative à la sphère*, Nouvelles Annales de Mathématiques, t. XII (1853), p. 433.

[18] O. Bonnet, *Mémoire sur la théorie des surfaces applicables*, J. Ec. Polyt. 42 (1867), p. 72–92.

[19] J.-B. Bost, *Introduction to compact Riemann surfaces, Jacobians, and Abelian varieties*, in *From Number Theory to Physics*, M. Waldschmidt, P. Moussa, J.-M. Luck, C. Itzykson, ed., Springer-Verlag, 1992.

[20] R. Bryant, *A duality theorem for Willmore surfaces*, J. Differential Geometry 20 (1984), p. 23–53. *Surfaces in conformal geometry*, Proc. Sympos. Pure Maths. Amer. Math. Soc. 48 (1988), p. 227–240.

[21] F. Burstall, D. Ferus, F, Pedit, U. Pinkall, *Harmonic tori in symmetric spaces and commuting Hamiltonian systems on loop algebras*, Ann. of Maths (II) 138 (1993), 173–212.

[22] F. Burstall, F. Pedit, *Harmonic maps via Adler-Kostant-Symes theory*, in [43].

[23] F. Burstall, J.H. Rawnsley, *Twistor theory for Riemannian symmetric spaces*, Lect. Notes in Math. 1424, Springer-Verlag, Berlin, 1990.

[24] F. Burstall, J.C. Wood, *The construction of harmonic maps into complex Grasmannians*, J. Differential Geometry 23 (1986), p. 255–297.

[25] E. Calabi, *Minimal immersions of surfaces in Euclidean spheres*, J. Differential Geometry 1 (1967), p. 111–125.

[26] G. Darboux, *Leçons sur la théorie générale des surfaces*, livre VII, Chapitre X, Gauthiers-Villars 1884.

[27] H. Dobriner, *Über die Flächen mit einem System sphärischer Krümmungslinien*, J. reine angew. Math. (Crelle's Journal) 94 (1883), p. 116–161.

[28] H. Dobriner, *Die Flächen Constanter Krümmung mit einem System Sphärischer Krümmungslinien dargestellt mit Hilfe von Theta Functionen Zweier Variabeln*, Acta Math. 9 (1886), p. 73–104.

[29] J. Dorfmeister, G. Haak, *Meromorphic potentials and smooth CMC surfaces*, preprint of the University of Kansas.

[30] J. Dorfmeister, F. Pedit, H.-Y. Wu, *Weierstrass type representation of harmonic maps into symmetric spaces*, Comm. in Analysis and Geom. 6 (1998), p. 633–668.

[31] B.A. Dubrovin, V.B. Matveev, S.P. Novikov, *Nonlinear equations of Korteweg-de Vries type, finite-zone linear operators, and Abelian varieties*, Russian Math. Surveys 31:1 (1976), p. 59–146.

[32] B.A. Doubrovine, S. Novikov, A. Fomenko, *Géométrie contemporaine*, édition Mir, Moscou, 3ème partie, Ch. 1, §12.

[33] J. Eells, L. Lemaire, *Another report on harmonic maps*, Bull. London Math. Soc. 20 (1988), p. 385–524.

[34] J. Eells, J.C. Wood, *Harmonic maps from surfaces to complex projective spaces*, Advances in Math. 49 (1983), p. 217–263.

[35] F. Ehlers, H. Knörrer, *An algebro-geometric interpretation of the Bäcklund transformation for the Korteweg-de Vries equation*, Comm. Math. Helv. 57 (1982), p. 1–10.

[36] L.P. Eisenhardt, *A treatise on the differential geometry of curves and surfaces*, Dover Publications, Inc., New York 1909.

[37] A. Enneper, *Analytisch-geometrische Untersuchungen*, Göttingen Nachr. (1868), p. 258–277 and p. 421–443.

[38] A. Enneper, *Untersuchungen über die Flächen mit planen und sphärischen Krümmungslinien*, Abh. Königl. Ges. Wissensch. Göttingen 23 (1878) and 26 (1880).

[39] A. Enneper, *Über die Flächen mit einem System sphärischen Krümmungslinien*, J. reine angew. Math. (Crelle's Journal) 94 (1883), p. 329–341.

[40] D. Ferus, F. Pedit, S^1-*equivariant minmal tori in* S^4 *and* S^1 *equivariant Willmore tori in* S^3, Math. Zeit. 204 (1990), 269–282.

[41] D. Ferus, F. Pedit, U. Pinkall, I. Sterling, *Minimal tori in* S^4, J. reine angew. Math. 429 (1992), 1–47.

[42] A. Fordy, *A historical introduction to solitons and Bäcklund transformations*, in [43].

[43] A. Fordy, J.C. Wood (editors), *Harmonic maps and integrable systems*, Aspects of Mathematics E23, Vieweg, 1994 (see also www.amsta.leeds.ac.uk/Pure/staff/wood/FordyWood/contents.html).

[44] C. Gardner, J. Greene, M. Kruskal et R. Miura, *Method for solving the Korteweg-de Vries equation*, Physic Rev. Lett. 19 (1967), p. 1095–1097.

[45] M. Giaquinta, G. Modica, J. Souček, *Cartesian currents in the calculus of variations*, vol. II, Springer Verlag 1998 (see Proposition 3 in Section 4.1.3).

[46] M. Guest, *Harmonic maps, loop groups and integrable systems*, Cambridge University Press, Cambridge 1997.

[47] M. Guest, Y. Ohnita, *Loop group actions on harmonic maps and their applications*, in [43].

[48] F. Hélein, *Applications harmoniques, lois de conservation et repères mobiles*, Diderot éditeur, Paris 1996; or *Harmonic maps, conservation laws and moving frames*, Diderot éditeur, Paris 1997. New edition: to appear in Cambridge University Press.

[49] F. Hélein, *Willmore immersions and loop groups*, J. Differential Geometry 50 (1998), p. 331–385.

[50] F. Hélein, *Weierstrass representation for Willmore surfaces*, in *Harmonic morphisms, harmonic maps, and related topics*, C. K. Anand, P. Baird, E. Loubeau and J. C. Wood, ed., Chapman and Hall/CRC Research Notes in Mathematics 413.

[51] F. Hélein, P. Romon, *Hamiltonian stationary Lagrangian surfaces in $\mathbb{C}^2$*, preprint math.DG/0009202, submitted to Comm. Analysis Geo.

[52] F. Hélein, P. Romon, *Weierstrass representation of Lagrangian surfaces in four-dimensional space using spinors and quaternions*, Comment. Math. Helv. 75 (2000), p. 668–680.

[53] F. Hélein, P. Romon, *Hamiltonian stationary Lagrangian surfaces in Hermitian symmetric spaces*, preprint math.DG/0010231.

[54] N. Hitchin, *Harmonic maps from a 2-torus to the 3-sphere*, J. Differential Geometry 31 (1990), p. 627–710.

[55] H. Hopf, *Uber Flächen mit einer Relation zwischen den Hauptkrümmungen*, Math. Nach. 4 (1950/1), p. 232–249.

[56] N. Kapouleas, *Compact constant mean curvature surfaces in Euclidean threespace*, J. Diff. Geometry 33 (1991), p. 683–715.

[57] K. Kenmotsu, *Weierstrass formula for surfaces of prescribed mean curvature*, Math. Ann. 245 (1979), p. 89–99.

[58] B. G. Konopelchenko, I.A. Taimanov, *Generalized Weierstrass formulae, soliton equation and Willmore surfaces I. Tori of revolution and the mKdV equation*, preprint.

[59] D.J. Korteweg, G. de Vries, *On the change of forms of long waves advancing in a rectangular canal, and a new type of long stationary waves*, Philos. Mag. Ser. 5, 39 (1895), p. 422–443.

[60] B. Kostant, *The principal three-dimensional subgroups and the Betti numbers of a complex simple Lie group*, Amer. J. Math. 81 (1959), p. 973–1032. *The solution to a generalized Toda lattice and representation theory*, Adv. Math. 34 (1979), p. 195–338.

[61] T. Kuen, *Flächen von constanten negativem Krümmungsmass nach L. Bianchi*, Sitzungsber. der Bayer. Akad. (1884).

[62] P.D. Lax, *Integrals of nonlinear equations of evolution and solitary waves*, Comm. Pure. Appl. Math. 21 (1968), p. 467–490.

[63] E. Lenz, *Über Enneperschen Flächen konstanten negativen Krümmungs-masses*, Dissertation, Universität Göttingen, 1879.

[64] S. Lie, *Zur Theorie des Flächen konstanter Krümmung*, Archiv for Mathematik og Naturvidenskab (I u. II) (1879) and (III) (1880), Kristinia.

[65] I. McIntosh, *Infinite dimensional Lie groups and the two-dimensional Toda lattice*, in [43].

[66] M. Melko, I. Sterling, *Integrable systems, harmonic maps and the classical theory of solitons*, in [43].

[67] U. Pinkall, *Hopf tori in S^3*, Invent. Math. 81 (1985), p. 379–386.

[68] U. Pinkall, I. Sterling, *On the classification of constant mean curvature tori*, Annals of Math. 130 (1989), p. 407–451.

[69] K. Pohlmeyer, *Integrable Hamiltonian systems and interactions through constraints*, Comm. Math. Phys. 46 (1976), p. 207–221.

[70] A. Pressley, G. Segal, *Loop groups*, Oxford Mathematical Monographs, Clarendon Press, Oxford 1986.

[71] C. Rogers, W.F. Shadwick, *Bäcklund transformations and their applications*, Mathematics in Science and Engineering, Vol. 161, 1982, Academic Press.

[72] E. Ruh, J. Vilms, *The tension field of the Gauss map*, Trans. Amer. Math. Soc. 149 (1970), p. 569–573.

[73] J. Russel, *Report on waves*, 14th Mtg. of the British Assoc. for the Advance. of Science, John Murray, London (1844), p. 311–390.

[74] M. Sato, *Soliton equations as dynamical systems on infinite dimensional Grassmannian manifolds*, RIMS Kokyuroku 439 (1981), p. 30–46.

[75] G. Segal, G. Wilson, *Loop groups and equations of KdV type*, Pub. Math. IHES, 61 (1985), p. 5–65.

[76] A. Sym, *Soliton surfaces and their applications (Soliton geometry from spectral problems)*, Lect. Notes in Physics 239 (1985), p. 154–231.

[77] W. Symes, *Systems of Toda type, inverse spectral problems and representation theory*, Invent. Math. 159 (1980), p. 13–51.

[78] I. A. Taimanov, *Modified Novikov-Veselov equation and differential geometry of surfaces*, Translations of the Amer. Math. Soc., Ser. 2, V. 179 (1997), p. 133–151.

[79] I. A. Taimanov, *Surfaces of revolution in terms of solitons*, Ann. Global Anal. Geom. 15 (1997), p. 419–435.

[80] I. A. Taimanov, *Finite gap solutions to the modified Novikov-Veselov equations, their spectral properties and applications*, Siberian Math. Journal 40 (1999), p. 1146–1156.

[81] G. Thomsen, *Über konforme Geometrie I: Grundlagen der konformen Flächentheoremrie*, Abh. Math. Sem. Hamburg (1923), p. 31–56.

[82] K. Uhlenbeck, *Harmonic maps into Lie groups*, Journal of Differential Geometry 30 (1989), 1–50.

[83] M. Voretzsch, *Untersuchung einer speziellen Fläche constanter mittlerer Krümmung bei welcher die eine der beiden Schaaren der Krümmungslinien von ebenen Curven gebildet wirdt*, Dissertation, Universität Göttingen (1883).

[84] R. Walter, *Zum H-satz von H. Hopf*, preprint, Universität Dortmund (1982).

[85] R. Walter, *Explicit examples to the H-problem of Heinz Hopf*, Geometriae Dedicata 23 (1987), p. 187–213.

[86] H. Wente, *Countereample to a conjecture of H. Hopf*, Pacific J. Math. 121 (1986), p. 193–243.

[87] H. Wente, *The capillary problem for an infinite trough*, Calc. Var. 3 (1995), p. 155–192.

[88] T. Willmore, *Riemannian geometry*, Oxford University Press, 1993.

[89] J.C. Wood, *The explicit construction and parametrization of all harmonic maps from the two-sphere to a complex Grasmannian*, J. reine angew. Math. 386 (1988), p. 1–31.

[90] J.C. Wood, *Harmonic maps into symmetric spaces and integrable systems*, in [43].

[91] V.E. Zakharov, A.V. Mikhailov, *Relativistically invariant two-dimensional models of field theory which are integrable by means of the inverse scattering problem method*, Soviet. Phys. JETP 47 (1978), p. 1017–1027.

[92] V.E. Zhakarov, A.B. Shabat, *Integration of nonlinear equations of mathematical physics by the method of inverse scattering*, Funktsional'nyi Analiz i Ego Prilozheniya, Vol. 13, No. 3, p. 13–22, July-September 1979.